# International Conference on Applied Sciences

# (ICAS 2013)



IOP Conference Series: Materials Science and Engineering Volume 57

## Wuhan, China
## 26 – 27 October 2013

ISBN: 978-1-63439-868-8
ISSN: 1757-8981

**Printed from e-media with permission by:**

Curran Associates, Inc.
57 Morehouse Lane
Red Hook, NY 12571

**Some format issues inherent in the e-media version may also appear in this print version.**



Printed by Curran Associates, Inc. (2015)



Institute of Physics
Dirac House, Temple Back
Bristol BS1 6BE UK

Phone: 44 1 17 929 7481
Fax: 44 1 17 920 0979

techtracking@iop.org

**Additional copies of this publication are available from:**

Curran Associates, Inc.
57 Morehouse Lane
Red Hook, NY 12571 USA
Phone: 845-758-0400
Fax: 845-758-2634
Email: curran@proceedings.com
Web: www.proceedings.com

# TABLE OF CONTENTS

IOP Conf. Series: Materials Science and Engineering **57** (2014) 011001 doi:10.1088/1757-899X/57/1/011001

# Preface

## International Conference on Applied Sciences (ICAS2013)

**Wuhan, P R China, 26–27 October, 2013**

Organizers:

*Military Economics Academy of Wuhan*

*"Politehnica" University of Timişoara*

in cooperation with

*Scientific Interdisciplinary Values for Education and Research Society (SCIVERS)*

*Wuhan University*

*Huazhong University of Science and Technology*

*University of Medicine and Pharmacy "Victor Babeş" Timişoara*

IOP Conf. Series: Materials Science and Engineering **57** (2014) 011001 doi:10.1088/1757-899X/57/1/011001

The **International Conference on Applied Sciences (ICAS2013)** took place in Wuhan, P R China from 26–27 October 2013 at the Military Economics Academy. The conference is regularly organized, alternately in Romania and in P R China, by "Politehnica" University of Timişoara, Romania, and Military Economics Academy of Wuhan, P R China, with the aim to serve as a platform for the exchange of information between various areas of applied sciences, and to promote the communication between the scientists of different nations, countries and continents. The conference has been organized for the first time in 15–16 June 2012 at the Engineering Faculty of Hunedoara, Romania.

The topics of the conference covered a comprehensive spectrum of issues:

1. Economical sciences
2. Engineering sciences
3. Fundamental sciences
4. Medical sciences

The conference gathered qualified researchers whose expertise can be used to develop new engineering knowledge that has applicability potential in economics, defense, medicine, etc. The number of registered participants was nearly 90 from 5 countries. During the two days of the conference 4 invited and 36 oral talks were delivered. A few of the speakers deserve a special mention: Mircea Octavian Popoviciu, Academy of Romanian Scientist — Timişoara Branch, *Correlations between mechanical properties and cavitation erosion resistance for stainless steels with 12% chromium and variable contents of nickel*; Carmen Eleonora Hărău, "Politehnica" University of Timişoara, *SWOT analysis of Romania's integration in EU*; Ding Hui, Military Economics Academy of Wuhan, *Design and engineering analysis of material procurement mobile operation platform*; Serban Rosu, University of Medicine and Pharmacy "Victor Babeş" Timişoara, *Cervical and facial infections — a real life threat*, among others.

Based on the work presented at the conference, 14 selected papers are included in this volume of IOP Conference Series: Materials Science and Engineering. These papers present new researches in the various fields of materials engineering, mechanical engineering, computers engineering, mathematical engineering and clinical engineering. It's our great pleasure to present this volume of IOP Conference Series: Materials Science and Engineering to the scientific community to promote further researches in these areas. We sincerely hope that the papers published in this volume will contribute to the advancement of knowledge in the respective fields.

All papers published in this volume of IOP Conference Series: Materials Science and Engineering (MSE) have been peer reviewed through processes administered by the editors of the ICAS2013 proceedings, Ludovic Dan Lemle and Yiwen Jiang. Special thanks should be directed to the organizing committee for their tremendous efforts in organizing the conference:

*General Chair*
Zhou Laixin, Military Economics Academy of Wuhan

*Co-chairs*
Du Qifa, Military Economics Academy of Wuhan
Serban Viorel-Aurel, "Politehnica" University of Timişoara
Fen Youmei, Wuhan University
Lin Pinghua, Huazhong University of Science and Technology

*Members*
Lin Darong, Military Economics Academy of Wuhan
Guo Zhonghou, Military Economics Academy of Wuhan
Sun Honghong, Military Economics Academy of Wuhan
Liu Dong, Military Economics Academy of Wuhan

We thank the authors for their contributions and we would also like to express our gratitude everyone who contributed to this conference, especially for the generous support of the sponsor:

HUNEDOARA MicroMega

*S C Micro-Mega HD S A*

*Ludovic Dan Lemle and Yiwen Jiang*
Coordinators of the Scientific Committee of ICAS2013

The members of the scientific committee who selected and reviewed the papers included in the Proceedings of ICAS2013 are:

IOP Conf. Series: Materials Science and Engineering **57** (2014) 011002 doi:10.1088/1757-899X/57/1/011002

# Peer review statement

All papers published in this volume of *IOP Conference Series: Materials Science and Engineering* have been peer reviewed through processes administered by the proceedings Editors. Reviews were conducted by expert referees to the professional and scientific standards expected of a proceedings journal published by IOP Publishing.

# Contribution on the influence of steel ladle processing (LF) upon the nitrogen removal rate

**E Ardelean[1], M Ardelean[1], A Josan[1] and C Pinca-Bretotean[1]**

[1] Engineering and Management Department, Polytechnic University of Timisoara, 5 Revolutiei Street, Hunedoara, postal code 331128, Romania

E-mail: erika.ardelean@fih.upt.ro

**Abstract.** Nitrogen, which is present in the composition of steel either dissolved or as a gas, represents an element, which is generally unwanted, except for the cases when the aim is to obtain nitrides or to increase the austenitic domain in stainless steels. The paper shows the results obtained in increasing the nitrogen removal rate during the secondary treatment of steel meant for oil industry pipes, into a Ladle Furnace-type installation. The processed data allowed the determination of variation domains, respectively graphical and analytical correlations between the nitrogen removal rate and the parameters of the secondary treatment process (bubbling duration, steel temperature and argon pressure).

## 1. Introduction

Nitrogen, similar to other gases, is undesirable in the liquid steel because it causes problems when it remains trapped in the solidified steel.

Nitrides, which are not removed, remain finely dispersed in the steel, where they serve as centers of crystallization, and unbound nitrogen that is found as nitrides in ferrite - together with oxygen - it increases the tendency to cold brittleness and aging.

In liquid steel, the nitrogen has partial solubility, which required its replacement as bubbling gas. However, with this technological limitation, nitrogen is present in the structure of steel, penetrating the liquid steel from the atmosphere of the steel making and secondary treatment department and/or casting department [1-3].

## 2. Industrial data

Data was taken from the steel pipe making process, on the followed technological flux: steel making (in EAF type EBT) secondary treatment (a LF facility). In order to determine the nitrogen content, it was sampled at steel evacuation in the ladle from the furnace, but in particular during the second treatment. In this way, there have been identified the argon bubbling parameters in LF facility (argon flow, bubbling duration, bubbling pressure) and thermal parameters, respectively the way is modified steel temperature during secondary treatment. It was intended a temperature at four different times: at the beginning, 20 minutes, 40 minutes from treatment start and final temperatures of steel [1].

All this data (considered as independent parameters) are presented in the following figures and analytical dependences, depending on the nitrogen removal efficiency.

IOP Conf. Series: Materials Science and Engineering **57** (2014) 012001 doi:10.1088/1757-899X/57/1/012001

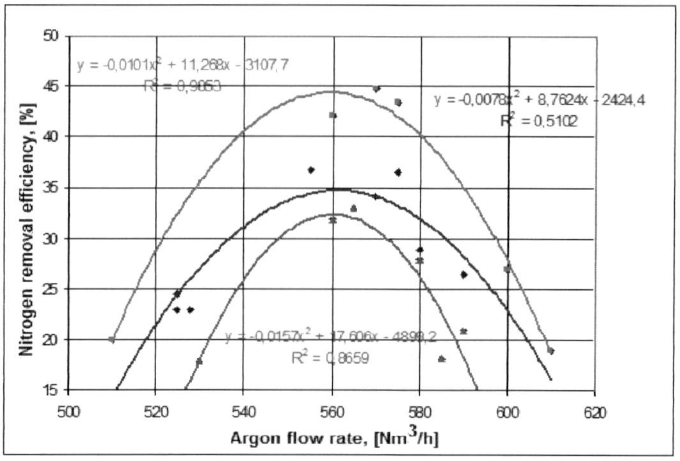


**Figure 1.** The influence of the argon flow on nitrogen removal efficiency

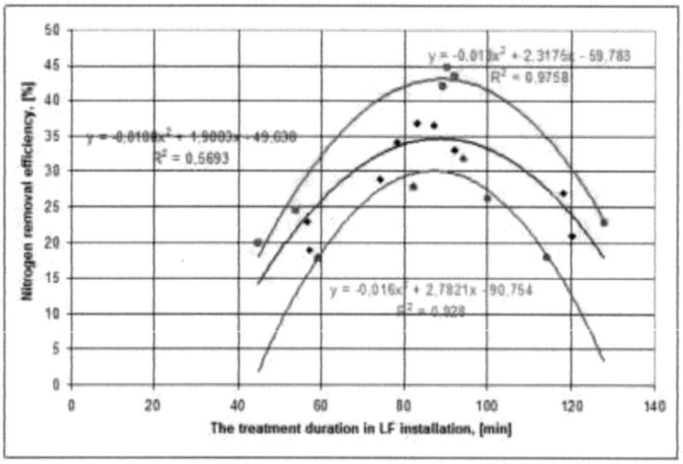


**Figure 2.** The influence of the bubbling duration on nitrogen removal efficiency

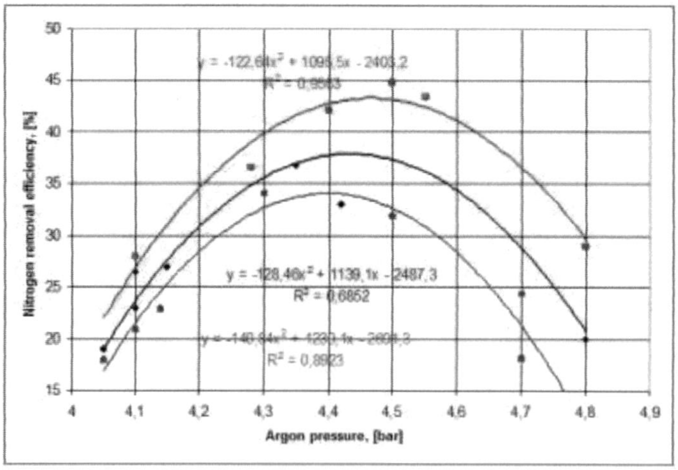


**Figure 3.** The influence of the argon pressure on nitrogen removal efficiency

It should be noted that, during the secondary treatment, liquid steel temperature is adapted according to its variation, reflected in the temperatures analysis presented in Figure 4 - Figure 7. At the end of treatment, the steel temperature is within technological limits necessary for the following secondary treatment, continuous casting process respectively.

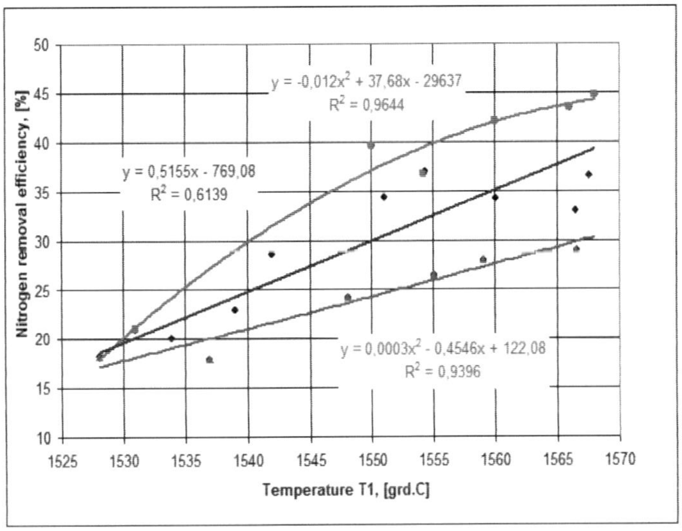


**Figure 4.** The influence of the steel temperature at the beginning of the secondary treatment on the nitrogen removal efficiency

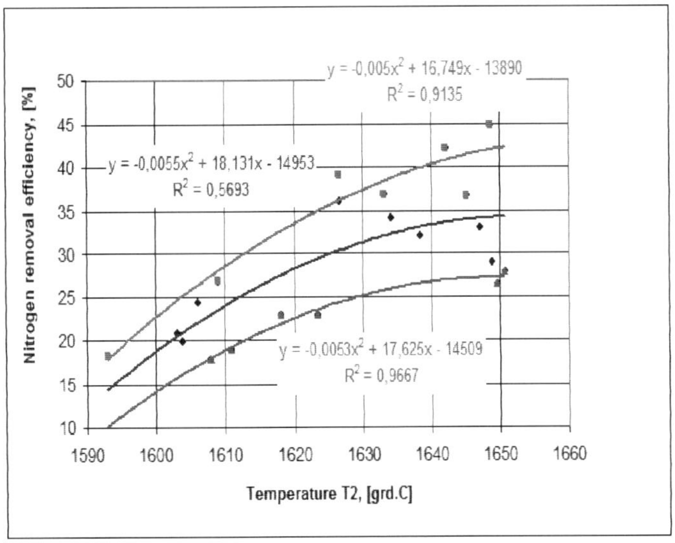


**Figure 5.** The influence of the steel temperature after 20 minutes from the start of secondary treatment on the nitrogen removal efficiency

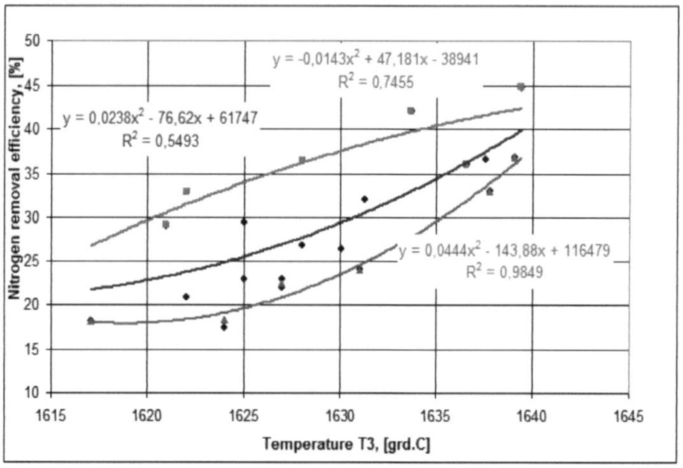


**Figure 6.** The influence of the steel temperature after 40 minutes from the start of secondary treatment on the nitrogen removal efficiency

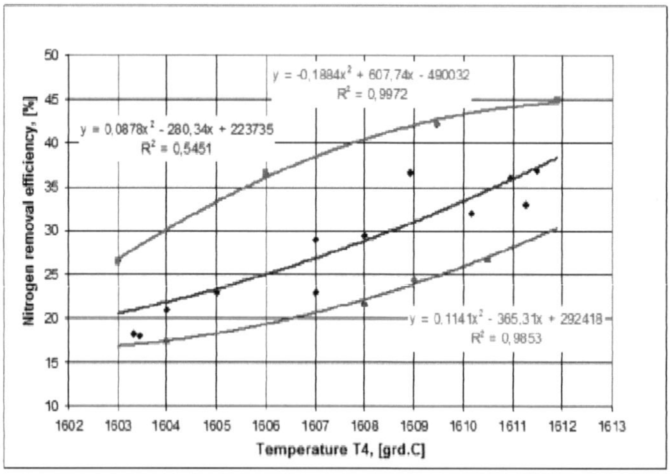


**Figure 7.** The influence of the steel temperature at the end of secondary treatment on the nitrogen removal efficiency

## 3. Discussions and results

Analysing the obtained results, presented in Figure 1-7, the following were found:

- for the correlation of the nitrogen removal efficiency depending on the argon flow at bubbling (Figure 1), the peak has the coordinates (561.69; 34.98) within the limits of the argon flow 525-595Nm$^3$/h, it leads to obtain a 25-35% nitrogen removal efficiency if we consider the correlation equation, or 25-45% if we take into account the correlation equation that is the upper limit of the range of variation;
- for the correlation of nitrogen removal efficiency depending on the bubbling (Figure 2), the peak has the coordinates (88.34; 34.99); the bubbling length is between 58-115 minutes, and it leads to a nitrogen removal from 25 to 35% if we take into account the correlation equation, or 25-45% if we take into account the correlation that is the upper limit of the range of variation;
- for the correlation of the nitrogen removal efficiency depending on the argon pressure at bubbling (Figure 3), the peak has the coordinates (4.43; 37.62); the argon pressure is between 4.1 to 4.75 bar and it leads to obtain a nitrogen removal yield between 23-35% if we consider the correlation

equation, or 23-45% if we take into account the correlation that is the upper limit of the range of variation;

- in what concerns the influence of the metal bath temperature on the nitrogen removal efficiency (Figure 4 - Figure 7), it is found that an increase of the temperature at the beginning and during the treatment, gives an increasing value for the dependent parameter; extreme points coordinates (minimum or maximum) are outside the technological domains, except for the correlation presented in Figure 5 that has the maximum point of coordinates (1648.18; 34.87);
- in terms of steel temperature, during processing in the LF facility, increasing its value determines the increase of the nitrogen removal efficiency at maximum; only one correlation has the point of maximum (for $T_2$=1648.18°C, $R_N$=34.87%) located at the upper limit of temperature variation domain, other two correlation have the minimum point ($T_3$ and $T_4$) located outside of temperature variation domain, below the lower limit of the domain and a correlation that is of the first degree).

## 4. Conclusions

Analysing the data and the discussions above, several conclusions can be drawn:

- from the mathematical point of view in all studied cases, the correlations are of the form of 2nd degree polynomial functions;
- for each case are presented the upper and lower variation limits, expressed both as analytical and graphical correlations;
- in three cases, the extreme point of the polynomial function is located within the technological limits, which from the technological point of view means that for the nitrogen removal efficiency a maximum value is reached for a given value of the technological independent parameter;
- for the nitrogen removal efficiency there were monitored possible values to be obtained, the independent technological parameters of the process being set into the limits required by the nitrogen removal efficiency;
- from the technological point of view, increasing the liquid steel temperature during the LF treatment, leads to a reduction in viscosity of the steel, which gives a higher diffusion of nitrogen from the steel in the argon bubbles, on the one hand, and on the other hand, a lower resistance to the movement of argon bubbles;
- increasing steel temperature must be within technological limits, exceeding the upper limit may cause a re-oxidation of the bath, or to a more intense absorption process than desorption process;
- framing temperatures during secondary treatment between technological limitations in hydrogen removal provides values for nitrogen removal efficiency within 25-45%.

## References

[1] Drăgoi F 2012 *Researches regarding the reduction of the gas content in steels developed and treated on the E.B.T – L.F. flow* (Timişoara: PhD Thesis, The Polytechnic University of Timisoara) pp 126

[2] Ştefănoiu R, Geantă V and Voiculescu I 2003 Modeling of the solid-liquid mass transfer in gas stirred systems *Proceedings of International Conference EUROMAT 2003 (Lausanne, Switzerland, 1-5 September 2003)* pp B3-1174

[3] Mândru C 2007 *Researches on improving the quality of customized steels by secondary metallurgy* (Bucureşti: PhD Thesis, Polytechnic University Bucureşti) pp 35

# Researches on the production of self-reducing briquettes from waste containing iron and carbon

**E Ardelean[1], M Ardelean[1], C Hărău[1] and E Crişan[1]**

[1] Engineering and Management Department, Politehnica University of Timisoara, 5 Revolutiei Street, Hunedoara, postal code 331128, Romania

E-mail: erika.ardelean@fih.upt.ro

**Abstract**. The extension of the raw material basis for the steel making industry represents a priority nowadays, within the context of sustainable development. The issue is of utmost importance particularly now when the environment legislation sets strict dumping conditions on the one hand and, on the other hand, when huge quantities of waste, already deposited in dumps and ponds raise serious problems in terms of meeting such conditions. The paper introduces some researches on exploiting waste containing iron and carbon in steel making; this powdery and small-grain waste is to be processed into briquettes. The self-reducing briquettes can be used in steel elaboration in electric arc furnaces, replacing the scrap, which is scarce.

## 1. Introduction

The iron oxide reduction takes place by direct way with carbon and indirect with carbon monoxide and hydrogen. In practice, it frequently occurs the situation in which the reduction is combined both with carbon (solid reducing agent) and with reducing gas (carbon monoxide and hydrogen).

Typically, iron ores are used in obtaining pig iron in blast furnace, either directly or in the form of pellets, briquettes, and agglomerate or sponge iron. In the electric arc furnaces, iron ore can be used as the oxidizing material introduced during oxidation period. However, due to intensive exploitation of the deposits, the quantities of ore are greatly diminished.

Furthermore, in the context of sustainable development, increasingly they focus more on recycling wastes containing elements which are useful in steel plants, waste with a high content of iron and carbon, but low in the content of unwanted elements. Thus, wastes from iron and steel industry can be used, such as blast furnace dust, EAF dust, scale from mills, sludge from blast-agglomeration, sideritic wastes; also there may be used any waste from other industries, but with adequate chemical composition: red mud (resulting from the alumina production), thermo-energetic ash, etc [1-3].

## 2. Laboratory experiments

The experiments took place in the laboratories of the Faculty of Engineering Hunedoara, Polytechnic University of Timisoara, and aimed to obtain compacted self-reducing briquettes from small and pulverous waste with high content of iron and carbon [1,2].

To obtain the briquettes, there were used the following waste categories: dust from furnaces and steel plants, rolling mills scale, red mud, graphite powder and for chemical composition correction and for bonding, we used lime and bentonite. The briquettes chemical composition is presented in table.1.

In these experiments, we performed briquettes reducing at a temperature range of 500-1300$^0$C, for a period of 22 minutes (average time). Experiments were made in a particular setting similar to the Hamborn furnace, and, in a certain degree, similar of FASMET process [4].

**Table 1.** Briquettes chemical composition, subjected to reduction processes

| No. recipe | Recipes chemical composition, [%] | | | | | | | | | | |
|---|---|---|---|---|---|---|---|---|---|---|---|
| | $FeO^*$ | $Fe_2O_3^*$ | C | $SiO_2$ | $Al_2O_3$ | CaO | MgO | MnO | S | $P_2O_5$ | other oxides |
| 1 | 10.402 | 52.951 | 19.09 | 6.421 | 2.432 | 5.81 | 0.58 | 1.224 | 0.489 | 0.129 | 1.026 |

*$Fe_{total}$ = 45.28%

For heating there have been used two resistance furnaces to reduce in a certain period of time several briquettes, on the one hand, and on the other, to provide charging the melting furnace with hot briquettes (newly reduced). In the first furnace (it may be heated up to 1200$^0$C) has been heated to a temperature from 650$^0$C to 1050$^0$C and in the second furnace from 1050$^0$C to 1300$^0$C (it may be heated up to 1700$^0$C). After a maintaining at temperature of 1050$^0$C for a period of 15-16min., the briquettes were removed and placed in the second furnace where, after final heating, they were maintained for 6-7min. to a temperature of 1250-1300$^0$C.

In the first furnace 24 briquettes can be heated at the same time, while in the second furnace can be heated only 4 briquettes. The total time of these operations was 115min (including times for transfer the briquettes in the other furnace). Randomly it was determined the metallization degree for 4 briquettes (equation (1)), the average of which is 93.48%, being considered for future technological calculations.

$$g = \frac{\%Me_{reduced}}{\%Me_{total}} 100, [\%] \quad (1.)$$

Hot briquettes were introduced in Tamman furnace for melting. The Tamman crucible furnace has a capacity of 3kg cast iron, so, given the iron content of the briquettes and their specific weight, in the crucible was charged as melting 2.75kg briquettes. To obtain a slag basicity ratio $I_B = CaO/SiO_2$ = 1.15-1.30 we made the addition of lime to 0.03 – 0.04kg.

Because the melting process has taken place in the graphite crucible, the carburization (with carbon from graphite) leads to advanced usage; so it has been made and addition of coke of 0.040kg/charge

In the conducted experiments, we have produced a number of 3 charges, in Tables 2 and 3 being presented the chemical composition of cast iron and slag.

The melting time was 30 -35min.; when melting has been completed the molten metal (cast iron) and slag have been cast in the graphite ingot. After that it has been determined the weight of cast iron and slag, the metal removal (yield) have been presented in Table 2.

**Table 2.** Cast irons chemical composition

| Charge no. | Cast irons chemical composition, [%] | | | | | | Yield, [%] |
|---|---|---|---|---|---|---|---|
| | C | Mn | Si | P | S | Fe | |
| 1 | 3.68 | 0.78 | 0.67 | 0.14 | 0.06 | 93.77 | 49.13 |
| 2 | 3.95 | 0.88 | 0.79 | 0.16 | 0.05 | 94.17 | 48.41 |
| 3 | 3.84 | 0.92 | 0.72 | 0.16 | 0.05 | 94.31 | 48.09 |

**Table 3.** Slags chemical composition

| Charge no. | Slags chemical composition, [%] | | | | | | | $CaO/SiO_2$ |
|---|---|---|---|---|---|---|---|---|
| | CaO | $SiO_2$ | $Al_2O_3$ | MgO | FeO | MnO | S | |
| 1 | 43.21 | 36.74 | 8.42 | 5.82 | 0.97 | 0.81 | 1.81 | 1.18 |
| 2 | 42.98 | 35.84 | 9.10 | 6.12 | 0.87 | 0.75 | 1.56 | 1.20 |
| 3 | 44.34 | 36.05 | 9.56 | 8.05 | 0.91 | 0.96 | 1.83 | 1.23 |

The second part of the experiments was performed in an induction furnace with a capacity of 10kg (cast iron /steel), with the metal bath volume of 1,45dm$^3$ (it was taken into account the molten metal that has a specific gravity equal to 6.9kg/dm$^3$ and 3kg/dm$^3$ for liquid slag). In the research it was considered that hot iron fills about 60% of the crucible volume and the rest returns to slag resulting from the reduction and smelting of the briquettes.

There have been developed 2 charges, in all cases the charges had 2 metal components (Table 4) as follows: 5kg solid cast iron + (3.0-3.5)kg briquette + 0.130kg lime (to provide a basicity ration $CaO/SiO_2 = 1.15$) + 0.100kg fluoride (to increase the fluidity of slag).

**Table 4.** Electric induction furnace load composition, [%]

| Charge no. | Charge components, [kg] | | | | Total, [kg] | Type briquettes |
|---|---|---|---|---|---|---|
| | cast iron | briquettes | lime | fluorine | | |
| 1 | 5 | 3.5 | 0.13 | 0.10 | 8.73 | Recipe 1 (table 1) reduced |
| 2 | 5 | 3.0 | 0.15 | 0.12 | 8.27 | Recipe 1 (table 1) nonreduced |

As mentioned above, in each charge its loaded into a furnace a solid cast iron and then the furnace was connected, and, as melting progressed, it was gradually introduced all of the cast iron. Because the charge was clean, the amount of resulting slag was practically insignificant (after cast iron smelting we extract average of 215g slag). When that melting was completed, the temperature was measured and it was sampled to determine the metal melt chemical composition (Table 5). After heating for 5 min., it has been continued the charging of the briquettes, thus ensuring the continuity of addition. Along with the addition of briquettes there were also made additions of lime and fluorine to correct the slag chemical composition.

**Table 5.** Cast iron chemical composition.

| Charge no. | Chemical composition, [%] | | | | | |
|---|---|---|---|---|---|---|
| | C | Mn | Si | P | S | Fe |
| 1 | 4.03 | 0.96 | 0.77 | 0.14 | 0.06 | 94.04 |
| 2 | 3.95 | 0.84 | 0.79 | 0.16 | 0.08 | 94.18 |

All the materials were weighed before charging into the furnace. After the completion of melting the entire furnace charging, it was maintained 5 minutes for thermal mixing and fluidization of slag, and then both the metal melt and the slag were poured in graphite ingots and were weighed after cooling with extracted slag, in order to perform mass balance.

Also for determining the recovery rate of iron from briquettes it was determined the chemical composition of cast iron and slag.

During the addition the following were determined:

*Charge No.1:*

- the addition of tubular briquettes, with a degree of metallization 93.48% it was in quantity of 2.46 kg briquettes, added as melting;
- the shape and porosity of metal briquettes favored their melting in relatively short time 9 min.

*Charge No.2*

- the addition of tubular self - reducing briquettes but initially not reduced was in the amount of 3.00kg added as melting;
- the tubular form of briquettes allow their rapid heating, which ensures intensive development of carbon reduction reactions, as demonstrated by the flames emitted from the surface of the metal bath, formed as a result of carbon monoxide combustion resulted from reduction reactions;
- briquettes time melting was 13 min.

Below is the presented the charge balance and results of laboratory tests, final tests on iron and slag, and technical calculations.

**Table 6.** Charge balance

| No. | Charge, [kg] | | | | | The melt, [kg] | | | | | Gases dust [kg] |
|---|---|---|---|---|---|---|---|---|---|---|---|
| | Cast iron | Briquettes | [1]Addition I | [2]Addition II | Total | Cast iron I | Slag I | Cast iron II | Slag II | [3]Total melt | |
| 1 | 5.0 | 2.46 | 0.230 | - | 7.69 | 4.786 | 0.214 | 5.791 | 1.57 | 7.575 | 0.12 |
| 2 | 5.0 | 3.0 | 0.230 | - | 8.23 | 4.797 | 0.203 | 5.707 | 1.31 | 7.22 | 0.41 |

[1] lime + fluorine; [2] ferrosilicium and graphite powder; [3] cast iron + slag I + slag II.

**Table 7.** Cast iron chemical composition

| Charge no. | Chemical composition, [%] | | | | | |
|---|---|---|---|---|---|---|
| | C | Mn | Si | P | S | Fe |
| 1 | 3.27 | 0.71 | 0.66 | 0.12 | 0.05 | 95.29 |
| 2 | 2.65 | 0.42 | 0.54 | 0.13 | 0.08 | 96.18 |

**Table 8.** Slag chemical composition

| Charge no. | Chemical composition, [%] | | | | | | | $CaO/SiO_2$ |
|---|---|---|---|---|---|---|---|---|
| | CaO | $SiO_2$ | $Al_2O_3$ | MgO | FeO | MnO | S | |
| 1 | 41.01 | 34.24 | 8.42 | 5.82 | 0.97 | 0.81 | 1.81 | 1.19 |
| 2 | 31.18 | 25.04 | 8.10 | 6.12 | 27.17 | 0.75 | 1.56 | 1.25 |

**Table 9.** Iron balance

| Charge no. | Iron, [kg] | | | | | $\eta_{rec.Fe}$, [%] |
|---|---|---|---|---|---|---|
| | $Fe_{F.I.}$ | $Fe_{F.II.}$ | $Fe_{F.I.} - Fe_{F.II}$ | $Fe_{briquettes}$ | $Fe_{slag}$ | |
| 1 | 4.5 | 5.518 | 1.018 | 1.104 | 0.086 | 92.21 |
| 2 | 4.518 | 5.489 | 0.971 | 1.346 | 0.375 | 72.13 |

1) The iron quantity from the cast iron before adding the briquettes
2) The iron quantity from the cast iron at 5 minutes after adding the briquettes

$$\eta_{rec.Fe} = \frac{Fe_{F.II.} - Fe_{F.I.}}{Fe_{Briquette}} \cdot 100, [\%] \qquad (2.)$$

where: $\eta_{rec.Fe}$ - iron recovery degree from briquettes;
$Fe_{F.I.}$- the quantity of iron in the cast iron before briquettes addition;
$Fe_{F.II.}$ - the quantity of iron in the cast iron at 5min. after briquettes addition.

## 3. Conclusions

From the above data, it appears that, through the melting of reduced briquettes in furnaces where conditions for the dissolution of carbon are created, we can obtain pig iron.

Slag resulting from the melting of the sterile from briquettes and the addition of lime, corresponds in terms of basic chemical composition to blast furnace slag.

Therefore it can be considered that in processing briquettes in Hamborn furnaces, they will be subject to the same physicochemical processes.

Analysing from the technological point of view, the following results were found:

- by melting reduced briquettes in furnaces, where conditions for dissolution of carbon into the steel bath are created, pig iron can be obtained;
- slag resulting from melting of the sterile and the addition of lime from briquettes is fluid, it has a low content of FeO and from the point of view of the chemical composition corresponds to blast furnace slag;
- at melting of briquettes in induction furnace where an iron bath was previously formed, different values for the iron recovery efficiency are obtained;

- the best values for the efficiency of iron recovery are obtained for the briquettes with a high degree of metallization, which can be explained by the fact that from their melting the slag that resulted had a lower oxidation capacity, on the one hand, and on the other slag has a lower fluidity;
- the lowest value for the iron recovery rate was obtained from melting of briquettes that are not initially reduced, condition caused by low slag flow resulting from the melting of the briquettes;
- the addition of the reducing mixture in furnace during the melting of briquettes lead to the increase of the iron recovery efficiency;
- given the variety of waste, both in terms of chemical composition and particle size, as well as small size waste beyond the smallest, but made in 250mm size, can be a viable technological solution for processing in Hamborn furnaces.

**References**

[1] Crisan E 2013 *Researches regarding recovery in siderurgy of small and pulverous waste containing iron and carbon* (Timişoara: PhD thesis, Politehnica University of Timişoara) pp 129

[2] Heput T, Socalici A, Ardelean E, Ardelean M, Constantin N and Buzduga R 2011 *Recovery of small and pulverous wastes* (Timişoara: Politehnica Publishing House) p 158

[3] Nicolae A and others 2004 *Sustainable development in the steel industry by capitalizing by-product materials* (Bucureşti: Printech Publishing House) p 428

[4] Hritac M, Hritac M, Iorga G, Savin D, Popescu L and Constantin N 2006 Framework methodology for determining the match between waste elements-processing technology, for the purpose of increasing the possibilities and level of usage of such waste *Proceedings of National Conference ROMAT (Bucuresti,Romania, 28-29 September 2006)* pp 389

International Conference on Applied Sciences (ICAS2013) IOP Publishing
IOP Conf. Series: Materials Science and Engineering **57** (2014) 012003 doi:10.1088/1757-899X/57/1/012003

# Contributions to steel semi-finished parts quality improvements

**A Socalici[1], E Popa[1], T Heput[1] and V Puțan[1]**

[1] Engineering and Management Department, Polytechnic University of Timisoara, 5 Revolutiei Street, Hunedoara, 331128, Romania

E-mail: virginia.socalici@fih.upt.ro

**Abstract**. The quality of finite steel products depends on steel content of hydrogen. On steel elaboration, in order to remove the dissolved hydrogen, at least one ladle secondary treatment is needed. The paper introduces the results obtained in the increase of hydrogen removal output, during steel secondary treatment inside the LF installation. Correlations were established between the hydrogen removal rate and the parameters of the secondary treatment (bubbling flow, bubbling time and bubbling pressures).

## 1. Introduction

Practically, during the usual elaboration, the hydrogen content of steel frequently varies between 1.8 – 10 ppm, and if the steel is the subject of the secondary treatment and vacuum casting metallurgy, the content of hydrogen is of maximum 1.8 – 2.2 ppm [1].

The main sources from which the hydrogen can turn into steel in liquid condition are: the metallic load and all the other materials used in the elaboration and casting, refractory materials, cooled elements of the oven and of installations for the treatment of liquid steel, humidity of the atmosphere in which one works, etc. The content of hydrogen must be limited in steel because it has negative effects on the semi-finished or on the finished products [4].

The negative influence of the hydrogen in steel can be observed in the following [2]:

- is one of the causes of breaths in steel ingots and castings of calm steels;
- contributes to the defect called "flakes" (tiny cracks) in steels alloyed with chromium and nickel, which substantially reduce the fatigue strength of steel parts;
- reduces the plasticity and tenacity of steel;
- affects the electrical and magnetic properties of the steels.

## 2. Experimental Research

The industrial experimentations regarding the influence of the slag characteristics on the hydrogen removal efficiency were performed on a technological flow of processing the steel made of an electric arc furnace, EBT type of 100 t capacity, LF installation and continuous cast installation with 5 wires.

At the secondary treatment of the steel in the LF installation the argon bubbling of the metallic bath takes place and also the addition for correcting the chemical composition and for deoxidation and desulphuration as well as the additions for reducing slag formation (lime, bauxite).The duration of the secondary treatment of the steel is 50-90 minutes, required by the timing with the continuous cast installation [3, 4]. To determine the hydrogen removal efficiency, samples of hydrogen were taken, before the insertion of the steel ladle in the LF installation and at the end of the treatment.

To underline a multiple influence of the technological agents on the hydrogen removal efficiency, I processed the data in the MATLAB programme and the results obtained are presented in a graphical manner. The surface regression equation (1nd and 2nd degree) regarding the hydrogen removal efficiency ($\eta_H$), bubbling pressures (Pb), bubbling flow (Db) and bubbling time (Tb) is:

$$\eta_H = -0.05 \cdot D_b + 12.43 \cdot P_b + 20.02 \quad (1)$$

Correlation coefficient: R = 0.38. Deviation from the regression surface: S = 7.76.

$$\eta_H = -0.005 \cdot D_b^2 + 0.02 \cdot D_b \cdot P_b - 80.02 \cdot P_b^2 + 5.59 \cdot D_b + 699.44 \cdot P_b - 3081.35 \quad (2)$$

Correlation coefficient R = 0.92. Deviation from the regression surface: S = 3.25.

Point of maximum: $P_b = 4.48$; $D_b = 555.39$; $\eta_H = 55.39$.

$$\eta_H = 13.7818 \cdot P_b + 0.11301 \cdot T_b - 24.3727 \quad (3)$$

Correlation coefficient: R = 0.44. Deviation from the regression surface: S1 = 7.51.

$$\eta_H = -66.55 \cdot P_b^2 + 0.088 \cdot P_b \cdot T_b - 0.009 \cdot T_b^2 + 589.37 \cdot P_b + 1.31 \cdot T_b - 1324.45 \quad (4)$$

Correlation coefficient: R = 0.95. Deviation from the regression surface: S = 2.49. Point of maximum: $P_b = 4.49$; $T_b = 88,12$; $\eta_H = 55.06$.

$$\eta_H = 0.09 \cdot T_b - 0.05 \cdot D_b + 68.11 \quad (5)$$

Correlation coefficient: R = 0.25. Deviation from the regression surface: S = 8.12.

$$\eta_H = -0.004 \cdot T_b^2 - 0.002 \cdot T_b \cdot D_b - 0.008 \cdot D_b^2 + 2.32 \cdot T_b + 9.41 \cdot D_b - 2688.05 \quad (6)$$

Correlation coefficient: R = 0.92. Deviation from the regression surface S = 3.25. Point of maximum: $T_b = 86,12$; $D_b = 560,42$; $\eta_H = 56,17$.

The surface regression is presented in Fig.1-6.

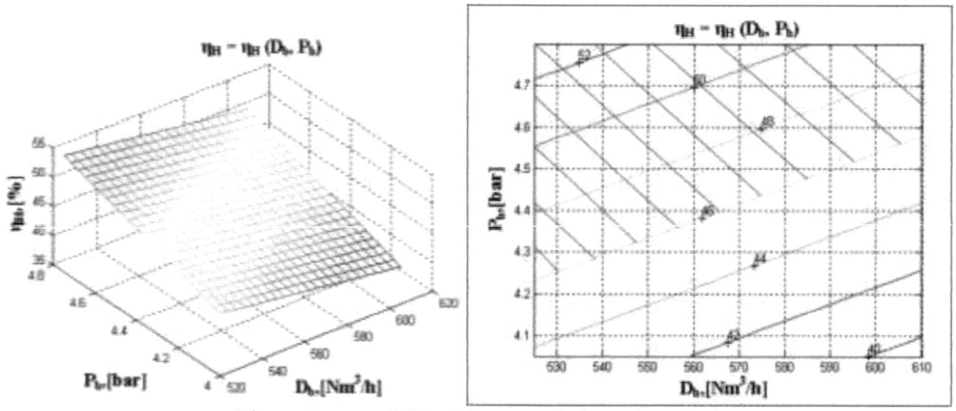


**Figure 1.** $\eta_H = f(D_b, P_b)$ – correlation of grade 1

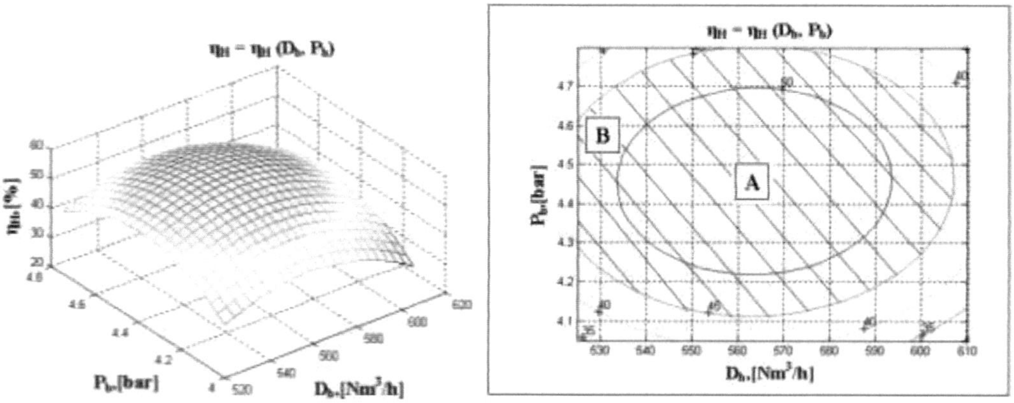


**Figure 2.** $\eta_H = f(D_b, P_b)$ – correlation of grade 2

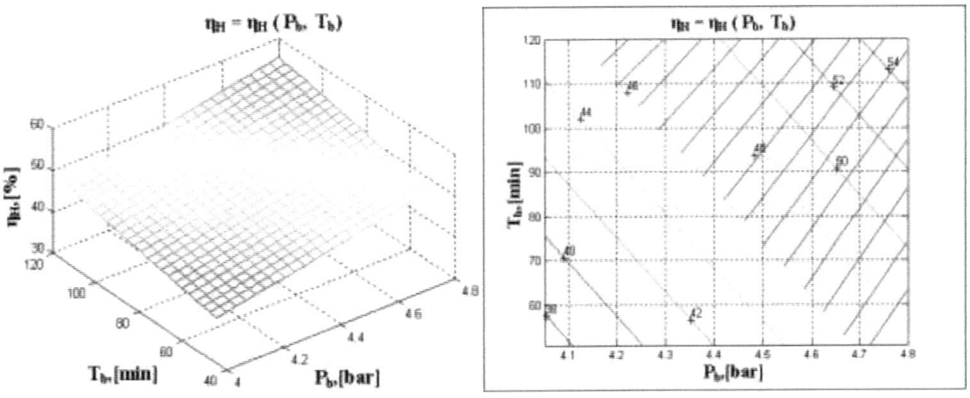


**Figure 3.** $\eta_H = f\ (T_b, P_b)$ – correlation of grade 1

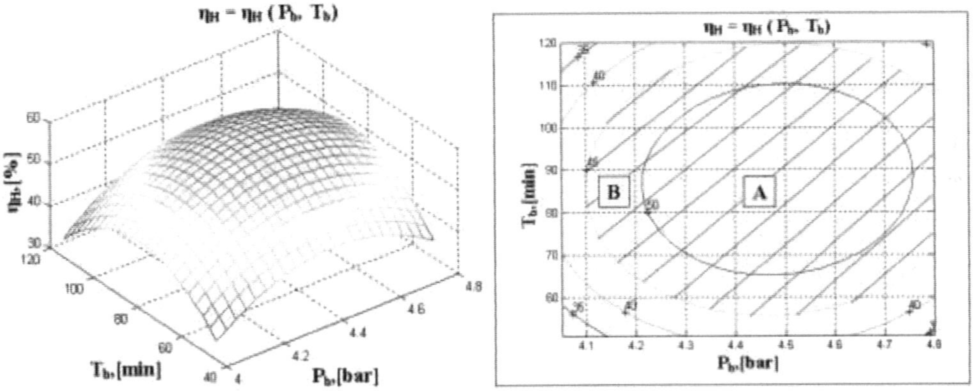


**Figure 4.** $\eta_H = f\ (T_b, P_b)$ – correlation of grade 2

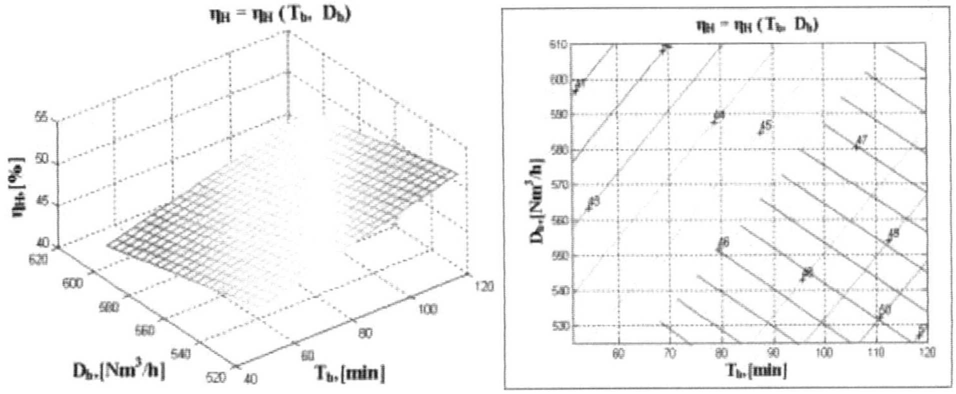


**Figure 5.** $\eta_H = f\ (T_b, D_b)$ – correlation of grade 1

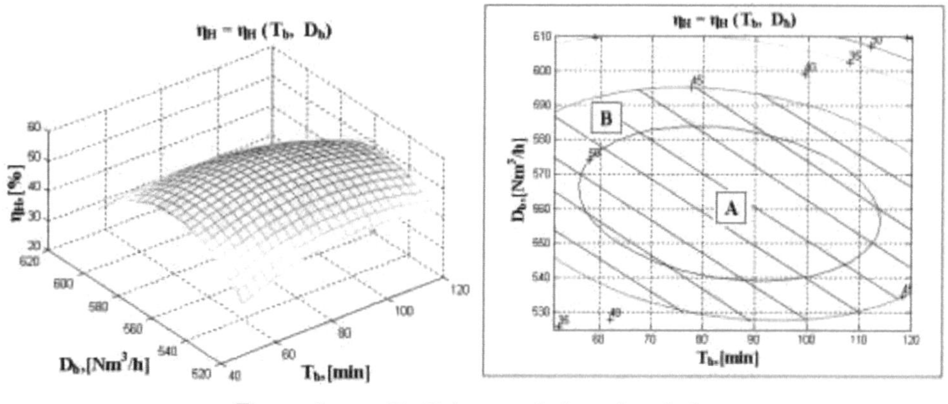


**Figure 6.** $\eta_H = (T_b, D_b)$ – correlation of grade 2

## 3. Conclusions

The analysis of the double correlations expressed analytically and graphically leads to the following conclusions:

- The variation of the independent parameters within technological limits determines for the dependent parameter a variation that also ranges within the technological parameters, which places it on a regression surface or in its vicinity, considering the standard dispersion, deviation and error;

- By intersecting the correlation surfaces with level planes, level curves were obtained, which allowed the determination of the variation limits for the independent parameters, in order to obtain a certain value for the dependent parameter. For each graphical representation, the sub-domains were shown (hatched A, B) for which the values for the dependent parameter should be found, which determines in fact the variation limits for the independent parameters;

- The analysis of the diagrams allows the establishment of the values for bubbling duration, argon flow rate and pressure on bubbling, so that the projection of the correlation surfaces should range within the hatched surface A, thus obtaining almost maximum values for the output of hydrogen removal.

## References

[1] Nica G, Socalici A, Ardelean E and Hepuţ T 2003 *Technologies to improve the quality of steel* (Timisoara: Mirton) p 48

[2] Hepuţ T, Ardelean E and Kiss I 2005 Some influence of the chemical composition upon the viscosity of synthetic slags used in continuous steel casting *Rev. Metal. Madrid* **41** 220

[3] Dragoi F, Socalici A, Hepuţ T and Popa E 2012 Researches on the influence of the slags formed in the instalation on the hydrogen removal efficiency *Rev. Metal. Madrid* **47** 477

[4] Dragoi F 2012 *Researches regarding the reduction of the gas content from the steels produced and treated on the EBT-LF technological flow* (Timisoara: PhD Thesis, Politehnica University of Timisoara) pp 85

IOP Conf. Series: Materials Science and Engineering **57** (2014) 012004 doi:10.1088/1757-899X/57/1/012004

# Contributions related to the control of steel ingot solidification

**A Socalici[1], E Popa[1], T Heput[1] and L Vîlceanu[1]**

[1] Engineering and Management Department, Polytechnic University of Timisoara, 5 Revolutiei Street, Hunedoara, postal code 331128, Romania

E-mail: virginia.socalici@fih.upt.ro

**Abstract**. The paper introduces the influence of the control over steel ingot solidification upon the quality characteristics of carbon steels. The laboratory experiments focused on the solidification control of steel ingots with cylindrical cross-section, by addition of crystallizing germs (micro-coolers) in their central area, in order to influence their inner structure and quality characteristics. The use of graphical and analytical correlations allows the establishing of optimal technological domains of variation for the steel casting parameters, with use of micro-coolers, in order to obtain the desired values for their mechanical characteristics.

## 1. Introduction

The basic problems that must be resolved casting steel it consist in obtain of homogeneous ingots from chemical point of view structural and mechanic properties. To mitigate the deficiency of ingots solidification in case of classical cast is necessary to adapt an efficient method to evacuate the heat from the steel in course of solidification. Usually the rapid heat evacuate is ensured through the creation of suddenly gradient of temperature in parts that solidification, as in liquid phase of the ingot. In this case, excepting the over heating must be field in most part and physic heat of cooling of the steel overlay that had been solidificated [1]. Conducting the crystalline structure can be realized using the temperature gradient (considerate basic parameter) from the liquid phase adjoining with the solidificated ingots font.

To conduct the solidification it can be utilized three methods, namely [2]:

- Using an alternative or continuous magnetic field for stopping the contraction of the currents from the liquid steel from the middle of the block of changing the movement direction
- Using the micro-coolers under forms of powder or granules metallic who reduce the gradient of temperature and influence the circulation of steel not solidified
- Enlarging the speed of solidification steel.

The micro-coolers used leads to reduce the piping extent, of the segregation, to improve the solidification structure and for values for physic-mechanic characteristic.

The method of ingot casting with micro-coolers is particularly recommended for the casting of large steel ingots, which are most likely to develop flaws such as: segregations, micro-shrinkages axial porosities and hot cracks. The process of steel crystallizing with exogenous germs introduced by means of micro-coolers differs considerably from the one used in common cast steel.

The micro-coolers introduced into the liquid alloy during casting, take over the overheat and also part of the solidification heat, some of them remaining into the liquid alloy as crystallizing germs. In order to obtain the desired effect, the micro-coolers have to be evenly distributed into the mass of liquid steel so that the action and sub-cooling areas created by each micro-cooler should partially overlap or be tangent.

The micro-coolers used to control the solidification process have to meet the following conditions [3]: to have high purity in terms of oxide inclusions, humidity should not exceed 0.25%, and not to have a stained surface; their chemical composition should be similar to that of the steel to be cast, and the grain size should be as even as possible.

## 2. Experimental researches

The steel was elaborated in the laboratory in a 10 kg induction furnace and the steel was cast into 2,5 kg circular cross-section ingots. The micro-coolers used were obtained by cutting pieces of steel wire with a diameter of 1mm. The micro-cooler addition was 1-4g/kg of cast steel. The micro-coolers were introduced into the mold when the latter was filled by 30%, 60% and 90% and in equal quantities (figure 1). These filling levels were established taking into consideration the local researches and the reference literature [4-7]. After having filled the mold, the end of the ingot was spread with anti-shrinkage dust. The resulting semi-finished parts were sampled in order to determine their physical and chemical characteristics.

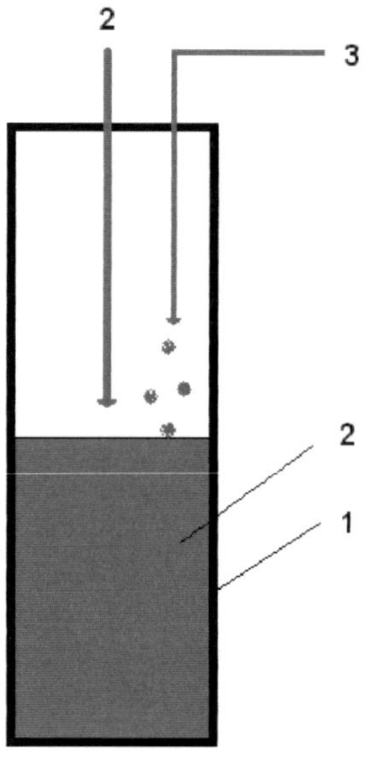


**Figure1.** Added into ingot mold micro-coolers:
1 – ingot; 2 – liquid steel; 3 – micro-coolers

The experimental data have been modeled by 1st, 2nd, 3rd and 4th degree polynomial surfaces. This polynomial modeling was done in Matlab. The coefficients of the regression surface for each degree of the polynomial surface were obtained by the method of the smallest squares.

The data resulting from the experiments are considered to be the values of independent variables *d* (diameter micro-coolers) and *q* (specific consumption of micro-coolers), and the values of dependent variable *A* (elongation) are compared to the values obtained by mathematical modeling, which gives the values of the deviations from the regression surface for each degree of the surface under analysis.

The equations describing the regression surfaces - 1st degree (*A*), the correlation coefficients and the deviation from the regression surface corresponding to this modeling are:

$$A = 1.10 \cdot d + 0.20 \cdot q + 21.97 \quad (1)$$

Correlation coefficient: R = 0.36. Deviation from the regression surface: S = 0.65.

Point of maximum: d = 1 mm; q = 4.0 g/kg; A = 23.88%.

The modeling resulted in the 1st degree polynomial surfaces shown in figure 2.

The equations describing the regression surfaces – 2nd degree (*A*), the correlation coefficients and the deviation from the regression surface corresponding to this modeling are:

$$A = -13.93 \cdot d^2 - 0{,}09 \cdot d \cdot q - 0.61 \cdot q^2 + 23.63 \cdot d + 3.32 \cdot q + 9.9452 \quad (2)$$

Correlation coefficient: R = 0.89. Deviation from the regression surface: S = 0.30. Point of maximum: d = 0.84mm; q = 2.65g/kg; A = 24,28%.

The modeling resulted in the 2nd degree polynomial surfaces shown in figure 3.

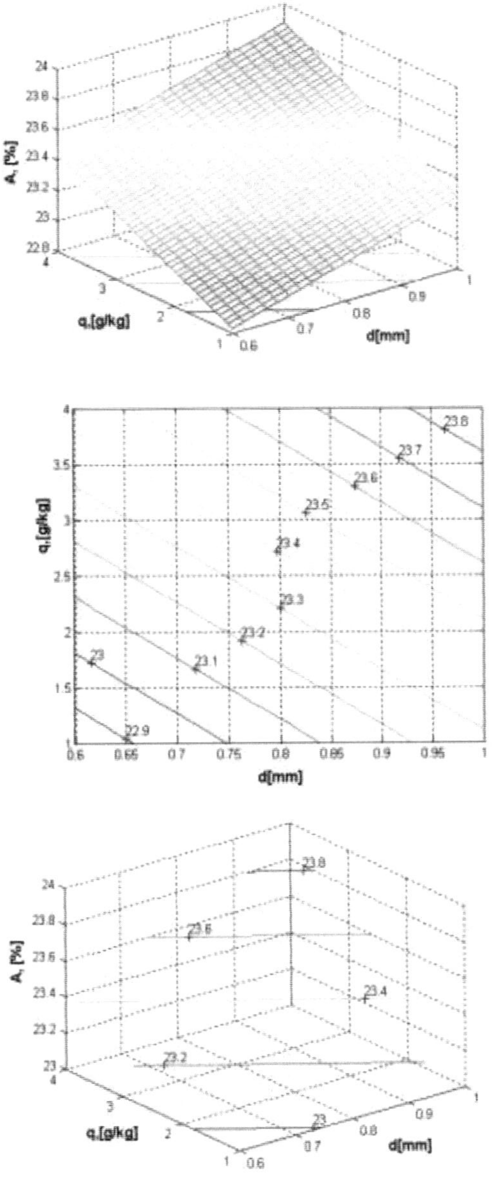


**Figure 2.** A=f(q,d) - the 1st degree polynomial surfaces

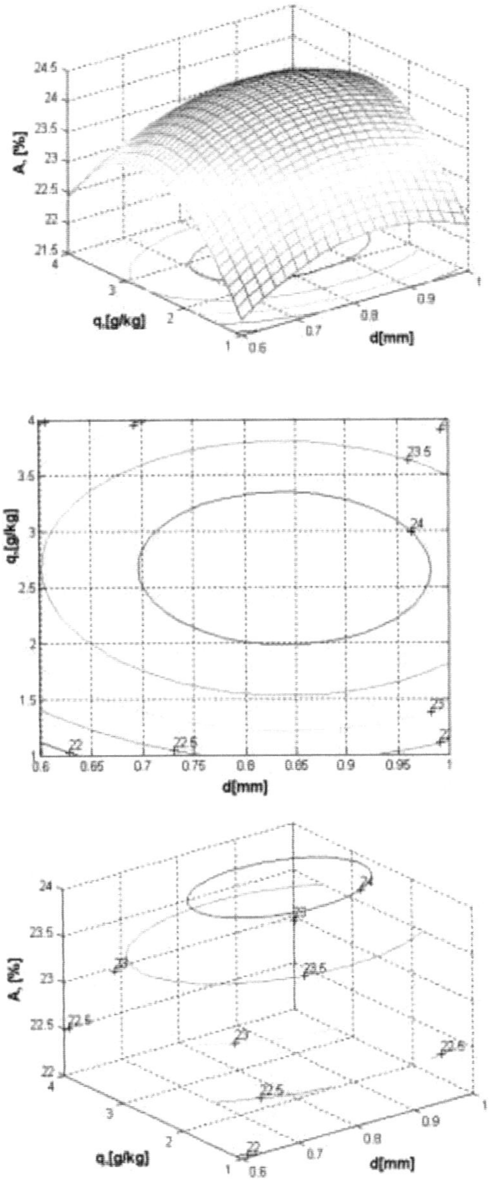


**Figure 3.** A=f(q,d) - the 2n degree polynomial surfaces

The equations describing the regression surfaces – 3rd degree ($A$), the correlation coefficients and the deviation from the regression surface corresponding to this modeling are:

$$A = -23.36 \cdot d^3 + 0.76 \cdot d^2 \cdot q + 0.25 \cdot d \cdot q^2 + 0.085 \cdot q^3 + 40.23 \cdot d^2 + -2.60 \cdot d \cdot q - 1.4563 \cdot q^2 - 16.0275 \cdot d + +6.2837 \cdot q + 18.0443 \quad (3)$$

Correlation coefficient: R = 0.90. Deviation from the regression surface: S = 0.26. Point of maximum: d = 0.86mm; q = 2.51g/kg; A = 24,29%.

The modeling resulted in the 3st degree polynomial surfaces shown in figure 4.

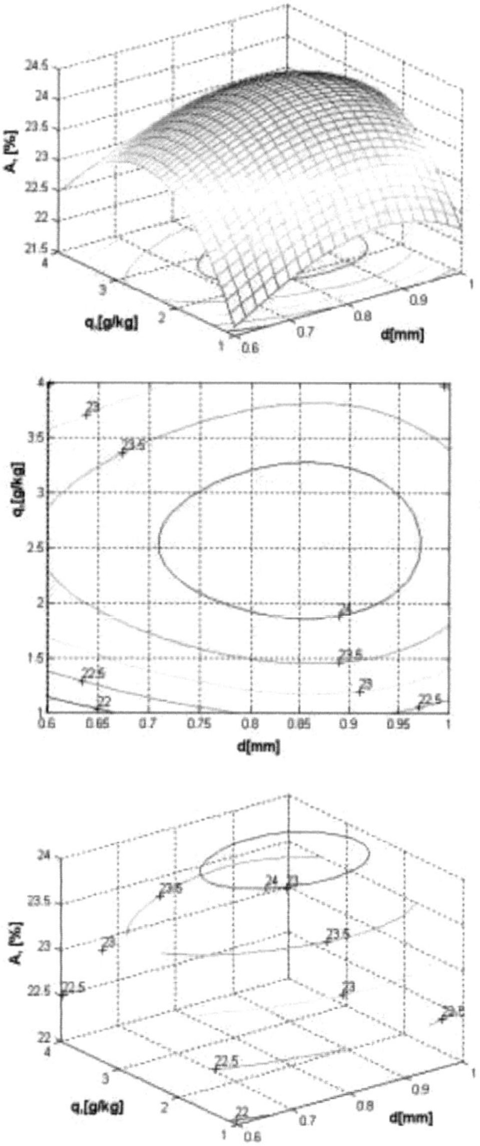


**Figure 4.** A=f(q,d) - the 1st degree polynomial surfaces

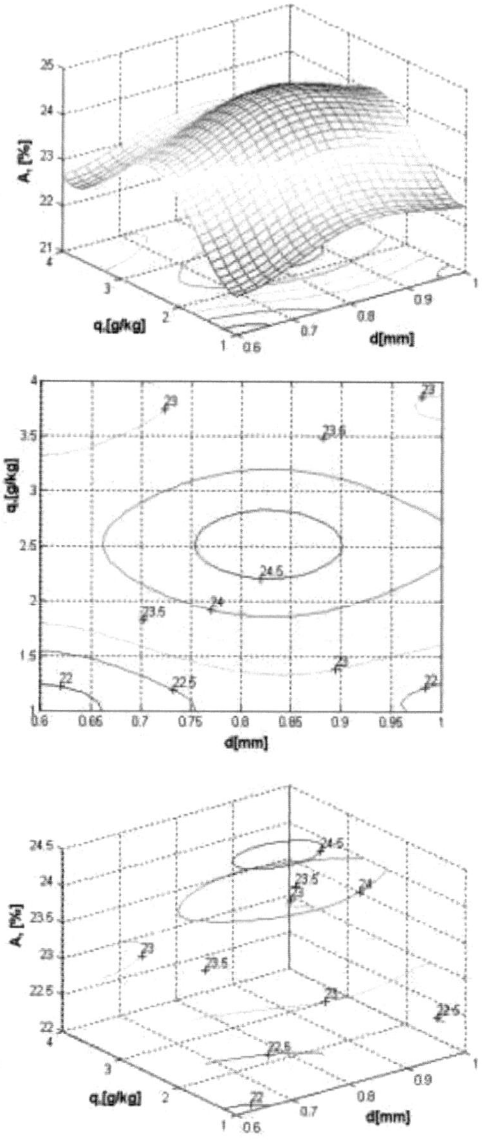


**Figure 5.** A=f(q,d) - the 2st degree polynomial surfaces

The equations describing the regression surfaces – 4th degree (*A*), the correlation coefficients and the deviation from the regression surface corresponding to this modeling are:

$$A = 322.85 \cdot d^4 + 3.73 \cdot d^3 \cdot q + 2.67 \cdot d^2 \cdot q^2 - 0{,}06 \cdot d \cdot q^3 + 0.42 \cdot q^4 - 1065.85 \cdot d \cdot q^3 - 21.60 \cdot d^2 \cdot q - 3.57 \cdot d \cdot q^2 - 4.10 \cdot q^3 + 1302.20 \cdot d^2 + 24.84 \cdot d \cdot q + 14.73 \cdot q^2 - 693.81 \cdot d - 24.44 \cdot q + 164.40 \quad (4)$$

Correlation coefficient: R = 0.95. Deviation from the regression surface: S = 0.19. Point of maximum: d = 0.82mm; q = 2.50g/kg; A = 24,66%.

The modeling resulted in the 4nd degree polynomial surfaces shown in figure 5.

## 3. Conclusions

From the analysis of correlations obtained by experimental data in Matlab program, and expressed in analytical and graphical form, results the following conclusions:

- were obtained equations of double correlation expressed by the polynomial function of degree 1 and 2, dependent parameters are the mechanical characteristics (A) and independent parameters of micro cooler diameter (d) and specific addition (q);
- the functions of second degree, provide great dependencies tracked;
- correlations allow establishing the optimal variation for diameter of the micro cooler and specific addition of them;
- graphics allow a relatively simple determination for a given value of the dependent parameter values for the independent parameters (the diameter of micro cooler and specific addition of them).

## References

[1] Eftimov V A 1976 *Steel casting and crystallization* (Bucureşti: Tehnica) p 209
[2] Bratu C and Sofroni L 1981 Conducting research on the crystallization process of steels using micro-coolers *Metalurgia* **6** 220
[3] Ioan R 2000 *Theoretical and experimental research in steel solidification* (Deva: Destin) p 70
[4] Andronache C, Socalici A, Hepuţ T and Popa E 2012 Research on the influence of steel ingot solidification process control on the tenacity characteristics *Met. Int.* **9** 243
[5] Andronache C, Socalici A and Hepuţ T 2013 The influence of micro-coolers on the physical-mechanical characteristics of the steel used in making railway monoblock wheels *Technical Gazette* **3** 419

# Design and engineering analysis of material procurement mobile operation platform

**H Ding**[1]**, J Li**[1]

[1] Military Economics Academy, Luojiadun 122#, Qiaokou District, Wuhan, Hubei, China

E-mail: dh192083@163.com

**Abstract**. The material procurement mobile operation platform (MPMOP) consists of six modules, including network operation, truck transportation, remote communication, satellite positioning, power supply and environment regulation. The MPMOP is designed to have six major functions, including online procurement, command control, remote communication, satellite positioning, information management and auxiliary decision. The paper implements an engineering analysis on the MPMOP from three aspects, including transportation transfinite, centroid, and power dissipation.

## 1. Introduction

In the process of material procurement, the geographical position of the customer could be changing and sometimes the material procurement institution is asked to follow them, and then the material procurement institution should be able to fulfill their procurement mission without fixed office and corresponding ancillary facilities. We design a kind of material procurement mobile operation platform (MPMOP), and the MPMOP is designed to be able to carry out the task of material procurement on the move according to the requirement of user. In this paper, we discuss the module of the MPMOP in part 2, the function of the MPMOP in part 3, and make an engineering analysis of the MPMOP in part 4.

## 2. Module

The MPMOP consists of six modules, including network operation, truck transportation, remote communication, satellite positioning, power supply and environment regulation.

### *2.1. Network Operation Module*

Having the function of network formation, information gathering, data processing and storage, real-time positioning, procurement scheme draft and decision, procurement document draft and printing, procurement information upload and download, the network operation module is the core of MPMOP. The network operation module consists of hardware system and software system. The hardware system includes server, microcomputer, switch, router, K-converter, synchronous/asynchronous modem, and laser printer. The software system provides the function of information management, resource analysis, auxiliary decision, network procurement, and so on.

### *2.2. Truck Transportation Module*
Consisting of chassis and car body, the truck transportation module mainly provides the function of moving and transportation. The application of off-road vehicle chassis let the MPMOP run faster and can adjust itself to different terrain. Having the function of shock absorption, electromagnetic shielding, and heat insulation, the car body includes door, window, annex, chair, ventilator and air-conditioner.

### *2.3. Remote Communication Module*
Consisting of wireless communication system and wire communication system, the remote communication module is mainly used to connect the MPMOP and the command network and the suppliers, and implement the transmission of data, audio and image. The wireless communication system consists of CDMA1X vehicle terminal, CDMA1X wireless network card, GSM wireless terminal, shortwave FH/adaptive terminal, shortwave modem, VHF FH radio set, and handheld walkie-talkie. The wire communication system consists of telephone, digital fax machine, intercom between car body and cab.

### *2.4. Satellite Positioning Module*
Using satellite navigation system, the satellite positioning module mainly provides the function of navigation and positioning, and can plan vehicle running route on the electric map.

### *2.5. Power Supply Module*
Consisting of external power supply input cable, vehicle-mounted electric generating set, vehicle battery, main vehicle generator, power supply box, and power distribution box, the power supply module can provide power support for other modules in four way, including civilian power supply, battery set power supply, generating set power supply, main vehicle generator power supply.

### *2.6. Environment Regulation Module*
Using overhead air-conditioner and adjustable vent, the environment regulation module can provide a comfortable work environment for the MPMOP.

## 3. Function
The MPMOP is designed to have six major functions, including online procurement, command control, remote communication, satellite positioning, information management and auxiliary decision.

### *3.1. Online Procurement Function*
The MPMOP includes internal resources database and external network. The internal resources database stores information of material procurement organization, material storage base and warehouse. Using wireless network technology, the MPMOP connects with local government procurement center and enterprise suppliers through Internet network, and implement online electronic procurement.

### *3.2. Command Control Function*
The MPMOP can acquire real-time information of material procurement quickly, make decision intelligently, manage efficiently, and implement material procurement at any area, any weather and any time.

### *3.3. Remote Communication Function*
The MPMOP is designed to have remote communication function, including wire communication, wireless communication and satellite communication, and then, the information of demand can be received timely and the instruction can be issued quickly.

### *3.4. Satellite Positioning Function*
The MPMOP needs to move to implement different material procurement mission. It must have the satellite positioning function so that the task allocation of material procurement and the material demand report can be more reasonable.

### *3.5. Information Management Function*
The MPMOP is designed to receive, process, analysis and transmit demand information and resource information of large number, and then, the MPMOP can make regional statistics of resource and analyze their potential.

### *3.6. Auxiliary Decision Function*
The MPMOP is designed to search resource database according to specific requirement and display these information in the electric map dynamically, and then help the commanders and staff make scientific decision quickly.

## 4. Engineering Analysis
In order to achieve the designed purpose, it is needed to make an engineering analysis on the MPMOP, and verify its ability to achieve the target. In this part, we implement an engineering analysis on the MPMOP from three aspects, including transportation transfinite, centroid, and power dissipation.

### *4.1. Analysis of Vehicle Transport Transfinite*
The purpose of vehicle transport transfinite analysis is to assure that the MPMOP satisfies the requirement of general transportation equipment, including railway transportation and highway transportation. The MPMOP installs side frame and main compartment on the IVECO NJ1056SHX6-T chassis, and makes the shape size as 6710 mm×2115 mm×3160 mm(length × width × height). According to GB1589, the shape size limit of 2 axle truck whose weight is between 3.5 tons and 8 tons is 8000 mm×2500 mm×4000 mm(length × width × height). According to GJB 2948, the shape size limit of material for railway transportation is 3000 mm × 3650 mm(width × height). The MPMOP satisfies the requirement of railway transportation and highway transportation.

### *4.2. Centroid Analysis of the MPMOP*
Because of the action of installing side frame and main compartment on IVECO NJ1056SHX6-T chassis, the centroid of the MPMOP must have changed. Centroid analysis is needed to assure the operation platform will drive safely. Take the center of front axle of the chassis as the origin of coordinate, the length of the MPMOP as the X axis, the width as the Y axis, the height as the Z axis, and the quality distribution condition can be seen clearly in Table 1.

*4.2.1. Analysis of the Centroid Position.* The total weight of the MPMOP is G=4935 kg. Take X as the vertical position of the MPMOP, then $X=(G_1\times X_1+G_2\times X_2+\cdots\cdots+G_n\times X_n)/G=2525.1$ mm. Take Y as the horizontal position of the MPMOP, then $Y=(G_1\times Y_1+G_2\times Y_2+\cdots\cdots+G_n\times Y_n)/G=3.3$ mm. Take Z as the height of the centroid, then $Z=(G_1\times Z_1+G_2\times Z_2+\cdots\cdots+G_n\times Z_n)/G=1199.2$ mm.

*4.2.2. Analysis of the Axle Load.* According to the requirement of the technical parameters of IVECO NJ1056SHX6-T, the load of the front axle must be less than 1550 kg, the load of the rear axle must be less than 3550 kg, the proportion of the load of the front axle to the total weight of the operation platform must be less than 36% and bigger than 20%, only then can the platform drive safely.

Take $G_f$ as the load of the front axle and $G_r$ the load of the rear axle, then $G_r=G\times X/L=4935\times 2525.1/3600=3461.5$ kg, $G_f=G-G_r=4935-3461.5=1473.5$ kg, and the proportion of the load of the front axle to the total weight of the platform $G_f/G$=1473.5/4935=29.9%. Conclusion can be made that the technical parameters of the MPMOP satisfies the requirement of safety and design.

**Table 1.** Quality Position Distribution Table

| Name of Project | Code | Quality (kg) | X (mm) | Y (mm) | Z (mm) |
|---|---|---|---|---|---|
| Chassis | $G_1$ | 2045 | 1050 | 0 | 670 |
| Car Body | $G_2$ | 1300 | 3550 | 0 | 1800 |
| Side Frame | $G_3$ | 300 | 3450 | 0 | 840 |
| Operating Personnel | $G_4$ | 300 | 3850 | 0 | 1428 |
| Equipment of Personnel | $G_5$ | 40 | 3850 | 0 | 1428 |
| Generating Set | $G_6$ | 70 | 5138 | 0 | 1340 |
| LCD TV & Installation | $G_7$ | 25 | 1540 | 0 | 2300 |
| Integrated Printer | $G_8$ | 10 | 2170 | -670 | 1820 |
| Digital Fax Machine | $G_9$ | 15 | 1730 | -670 | 1815 |
| Overhead Air-conditioner | $G_{10}$ | 68 | 3725 | 0 | 3040 |
| Mounting Frame of Outdoor Machine | $G_{11}$ | 15 | 1110 | -95 | 2600 |
| Ultra-short Wave Antenna | $G_{12}$ | 15 | 2820 | -500 | 2977 |
| Satellite TV Receiver | $G_{13}$ | 5 | 1888 | 500 | 2970 |
| Satellite User Receiver Antenna | $G_{14}$ | 5 | 4898 | 500 | 2965 |
| Left Rear Left Electric Cabinet | $G_{15}$ | 80 | 5093 | -698 | 2190 |
| Right Rear Left Electric Cabinet | $G_{16}$ | 80 | 5093 | 698 | 2190 |
| Front Left Electric Cabinet | $G_{17}$ | 60 | 1923 | -700 | 1315 |
| Notebook Computer | $G_{18}$ | 10 | 3258 | 0 | 1690 |
| Office Furniture | $G_{19}$ | 150 | 3605 | 0 | 1315 |
| Accessory Box | $G_{20}$ | 40 | 2500 | -850 | 686 |
| Cable Tray Box | $G_{21}$ | 70 | 2250 | 850 | 686 |
| Toolbox | $G_{22}$ | 30 | 4780 | 890 | 757 |
| UPS Battery | $G_{23}$ | 72 | 5143 | 650 | 1090 |
| Electric Fuel | $G_{24}$ | 80 | 3540 | 0 | 1340 |
| Others | $G_{25}$ | 50 | 3540 | 0 | 1140 |
| Total | G | 4935 | 2525.1 | 3.3 | 1199.2 |

*4.2.3. Analysis of the Left and Right Axle Load.* According to the technical parameters of the IVECO NJ1056SHX6-T chassis, the proportion of the difference of axle load between the left and the right part to the total weight ($\Delta G/G$) must be less than 3%, only then can the MPMOP drive safely. The horizontal position of the centroid Y=3.3 mm, the wheel spacing B=1683 mm, the load of the left part $G_L=(B/2-Y)\times G/B=838.2\times 4935/1683=2457.8$ kg, and the load of the right part $G_R=G-G_L=4935-2457.8=2477.2$ kg. The difference of axle load between the left part and the right part $\Delta G=G_L-G_R=19.4$ kg, and $\Delta G/G=0.39\%$.

*4.2.4. Lateral Stability Check Analysis.* According to the technical parameters of IVECO NJ1056SHX6-T chassis, the maximum lateral stability angle of the MPMOP $\beta_{max}$ must be bigger than 35°. Because $\tan\beta_{max}=B/2Z=1683/(2\times 1199.2)=0.71$ and $\beta_{max}=35.1° > 35°$, so the lateral stability satisfies the safety requirement.

*4.2.5. Longitudinal Stability Check Analysis.* According to the technical parameters of IVECO NJ1056SHX6-T chassis, the attachment coefficient ϕ between tire and road when the MPMOP is uphill and downhill must be bigger than 0.7. When the MPMOP is uphill, $\phi=(L-X)/Z=(3600-2525.1)/1199.2=0.9$. When the MPMOP is downhill, $\phi=X/Z=2525.1/1199.2=2.11$. L is the wheelbase of the MPMOP and L=3600mm. With the analysis, we can find that the attachment coefficient ϕ between tire and road when the MPMOP is uphill and downhill are both bigger than 0.7, and satisfy the requirement of longitudinal stability.

*4.2.6. Wind Load Stability Analysis.* The MPMOP is asked to be able to work even when the wind speed is level 8, so we take the maximum wind speed of level 8 20.7m/s as the input wind load when we analyze the wind load stability. The windward area of the side is biggest for the MPMOP and the support force of the tire is the smallest, so, we calculate the windward rollover torque M of the side of the MPMOP. Because the height of the MPMOP is less than 10 m while working, the MPMOP is mainly influenced by surface wind. The wind load $W_0$ is proportion to the square of the wind speed $V_0$, the coefficient of correspondence is 1600, and $V_0=20.7$ m/s, then $W_0=V_0^2/1600=0.27$ KN/m$^2$. The windward rollover torque M consist of car body frame's rollover torque $M_1$ and chassis's rollover torque $M_2$.

The measure of the side area of the car body frame $S_1=L\times W=4.115\times 2=8.23$ m$^2$. In the equation, L is the length of the car body frame, and W is the width of the car body frame. The wind load of the car body frame $F_1=W_0\times S_1=0.27\times 8.23=2.22$ KN. The equivalent point of this force is located at the center of the car body and the height $L_d=1917$ mm, then the car body frame's rollover torque $M_1=F_1\times L_d=2.22\times 1917=4256$ Nm.

The measure of the side area of the chassis can be achieved by the calculation of the closed contour of the MPMOP, then we get the measure of the side area of the chassis $S_2=4.4$ m$^2$, the height of the center of the closed contour h=745 mm, the wind load of the chassis $F_2=W_0\times S_2=0.27\times 4.4=1.19$ KN, the rollover torque of the chassis $M_2=F_2\times h=1.19\times 745=887$ Nm, the total of the rollover torque $M=M_1+M_2=4256+887=5143$ Nm.

The stability torque of the MPMOP $Mg=G\times(B/2+Y)\times 9.8/1000=40857$ Nm. In this equation, G is the total weight of the MPMOP and G=4935kg, B is the wheel spacing and B=1683 mm, Y is the horizontal position of the MPMOP and Y=3.3 mm. Because the rollover torque of the MPMOP M=5143 Nm is smaller than the stability torque Mg=40857 Nm, the wind load stability of the MPMOP satisfies the requirement of the safety.

### *4.3. Analysis of the Power Dissipation*

The following table shows the power dissipation of equipments of the MPMOP.

**Table 2.** Power Dissipation Statistical Table

| Serial Number | Equipment | Number | Maximum Power Dissipation(W) |
|---|---|---|---|
| 1 | VHF Radio | 1 | 120 |
| 2 | Ultra short Wave Modem | 1 | 10 |
| 3 | Cable Modem | 1 | 10 |
| 4 | Wired Intercom | 1 | 3 |
| 5 | Notebook Computer | 4 | 100 |
| 6 | Satellite Signal User Terminal | 1 | 30 |
| 7 | Digital Fax Machine | 1 | 20 |
| 8 | Video Conference System | 1 | 90 |

| | | | |
|---|---|---|---|
| 9 | Information Calendar | 1 | 5 |
| 10 | Power Distribution Control Cabinet Ventilator | 1 | 30 |
| 11 | Network System Accessories | 1 | 50 |
| 12 | PBX | 1 | 10 |
| 13 | Lighting | 4 | 100 |
| 14 | Multi-function Printing Machine | 1 | 300 |
| 15 | Air-conditioner | 1 | 1200 |
| 16 | Cabin Ventilator | 3 | 30 |
| 17 | Others | 1 | 115 |
| Total | | | 2883 |

According to the analysis of power dissipation of equipments of the MPMOP, a diesel generator of 3000W will satisfy the requirement of power consumption of all the equipments of the MPMOP.

## 5. Conclusion

According to the analysis above, we can make some conclusions as follows:

Conclusion 1: The MPMOP satisfies the requirement of railway transportation and highway transportation.

Conclusion 2: The technical parameters of the MPMOP, including the axle load, the left and right axle load, the lateral stability, the longitudinal stability and the wind load stability, satisfies the requirement of safety.

Conclusion 3: A diesel generator of 3000W will satisfy the requirement of power consumption of all the equipments of the MPMOP.

Conclusion 4: The MPMOP can fulfill its material procurement mission safely and efficiently.

## References

[1] Ahmad S and Schroeder R G 2001 The Impact of Electronic Data Interchange on Delivery Performance *Production and Operations Management* **10** 16

[2] Cachon G P and Fisher M 2000 Supply Chain Inventory Management and the Value of Information *Management Science* **46** 1032

[3] Chen M and Meixell M J 2003 Web Services Enabled Procurement in the Extended Enterprise: an Architectural design and Implementation *Journal of Electronic Commerce Research* **4** 140

[4] Sriram V and Banerjee S 1994 Electronic Data Interchange: Does Its Adoption Change Purchasing Policies and Procedurse? *International Journal of Purchasing and Materials Management* **30** 31

# Correlations between mechanical properties and cavitation erosion resistance for stainless steels with 12% Chromium and variable contents of Nickel

**I Bordeasu[1], M O Popoviciu[2], I Mitelea[1], B Ghiban[3], N Ghiban[3], M Sava[1], S T Duma[1] and R Badarau[1]**

[1] „Politehnica" University of Timisoara, Bvd. Mihai Viteazul No.1, 300222 Timisoara, Romania
[2] Academy of Romanian Scientists, Bvd. Mihai Viteazul No.1, 300222 Timisoara, Romania
[3] Politehnica University of Bucharest, Splaiul Independentei No.313, 060042 Bucharest, Romania

E-mail: ilarica59@gmail.com

**Abstract.** The running time of hydraulic machineries in cavitation conditions, especially blades and runners, depend on both chemical composition and mechanical properties of the used steels. The researches of the present paper have as goal to obtain new materials with improved behavior and reduced costs. There are given cavitation erosion results upon eight cast steels with martensite as principal structural constituent. The chromium content was maintained constant at approximate 12% but the nickel content was largely modified. The change of chemical content resulted in various proportions of austenite, martensite and ferrite and also in different cavitation erosion behavior. From the eight tested steels four have greater carbon content (approximately 0.1%) and the other four less carbon content (approximate 0.036%). All steels were tested separately in two laboratory facilities: T1 with magnetostrictive nickel tube (vibration amplitude 94 μm, vibration frequency 7000 ± 3% Hz, specimen diameter 14 mm and generator power 500 W) and T2 is respecting the ASTM G32-2010 Standard (vibration amplitude 50μm, vibration frequency 20000 ± 1% Hz, specimen diameter 15.8 mm and generator power 500 W). Analyzing the results it can be seen that the cavitation erosion is correlated with the mechanical properties in the way shown in 1960 by Hammitt and Garcia but is influenced by the structural constituents.

## 1. Introduction

An increased cavitation erosion resistance of materials is determined by high values of mechanical properties correlated with the chemical constitution, the nature and homogeneity of the structural constituents [1], [3], [7]. For hydraulic machinery details, running in industrial accepted cavitation condition, the use of materials with high cavitation erosion resistance is an important economic factor [2].

Significant results for the connection between cavitation erosion resistance and the material mechanical properties were obtained in the past by Garcia and Hammitt [5]. The mentioned relations which correlate one or many properties with the parameter 1/MDER (which define the cavitation erosion resistance) are practical power functions, whose exponent differ from a material to other,

IOP Conf. Series: Materials Science and Engineering **57** (2014) 012006 doi:10.1088/1757-899X/57/1/012006

depending also on the nature of the cavitation liquid. For example, Garcia considers that the increase of cavitation erosion resistance (1/MDER) is proportional to $HB^{1.8}$. The results of our researches obtained with a range of stainless steels having various structures (combinations between martensite, austenite and ferrite) show significant differences between the values of the scale and shape (power exponents) in the relations correlating the parameter 1/MDER with tensile strength ($R_m$), yield point ($R_{p0.2}$), Brinell hardness (HB) and ultimate resilience ($UR=(R_m)^2/2E$) in comparison with the value indicated by Garcia and Hammitt [5] which use test facilities similar to T1.

## 2. Tested materials

The tested materials are stainless cast steels used in manufacturing hydraulic turbines blades or pump impellers. In Tables 1 and 2 are presented the values of mechanical characteristics determined by Strength Materials Laboratories and the structural constituents, before cavitation exposure, obtained by using the Schäffler diagram [3], after computing chromium and nickel equivalents.

Because the tested steels are not standard brands, to simplify their identification it was conceived the notation presented in Table 1, which use the symbol of the principal chemical elements and the numbers showing the content, as follows:

- the chemical symbol of chromium (Cr) followed by the number 12, representing the percentage of this element, equal for all eight used steels;
- the chemical symbol of nickel (Ni) followed by a figure or a group of figures representing in percents, the content of this element;
- the chemical symbol of carbon (C) followed by the figure 1 or the number 036 representing the content of this element (either 0.1% C or 0.036% C).

**Table 1** Values of mechanical properties [6]

| **Steel** | **Carbon content %** | **Fracture strength $R_m$ [MPa]** | **Yielding strength $R_{p0,2}$ [MPa]** | **HRC (≅HB)** | **Necking, Z, %** | **Elongation $A_5$, %** |
|---|---|---|---|---|---|---|
| Cr12Ni0.5C1 | ≅ 0.1 | 1450 | 1020 | 44 (411) | 26.8 | 6.8 |
| Cr12Ni2C1 | | 1336 | 935 | 40 (369) | 25.4 | 6.9 |
| Cr12Ni6C1 | | 1540 | 1083 | 46 (434) | 24.8 | 6.3 |
| Cr12Ni10C1 | | 835 | 626 | 25 (254) | 32.2 | 11 |
| Cr12Ni2C036 | ≅ 0.036 | 968 | 678 | 29 (280) | 29.3 | 8.2 |
| Cr12Ni4C036 | | 989 | 695 | 30 (287) | 28.7 | 8 |
| Cr12Ni6C036 | | 1035 | 725 | 31 (297) | 30.1 | 8.7 |
| Cr12Ni8C036 | | 1002 | 701 | 30 (288) | 31.3 | 8.8 |

The data presented in Table 1 and 2 prove the diversity of the mechanical properties of the analyzed structures. It must be noted that for the steels with 0.1% of carbon, with the exception of Cr12Ni10C1, the mechanical properties $R_m$, $R_{p0.2}$ and HB have superior values in comparison to steels with 0.036% carbon.

**Table 2** Structural constituents [6]

| Steel | Chromiun Equivalent $Cr_{e,}$ [%] | Nickel Equivalent, $Ni_{e,}$ [%] | Structural constituent |
|---|---|---|---|
| Cr12Ni0.5C1 | 14.26 | 4.81 | 75% Martensite +25% Ferrite |
| Cr12Ni2C1 | 14.62 | 6.23 | 90% Martensite +10% Ferrite |
| Cr12Ni6C1 | 14.9 | 10.14 | 40% Martensite +60% Austenite |
| Cr12Ni10C1 | 14.66 | 14.74 | 100%Austenite |
| Cr12Ni2C036 | 13.27 | 3.15 | 55% Martensite +45% Ferrite |
| Cr12Ni4C036 | 13.1 | 5.25 | 86% Martensite +14% Ferrite |
| Cr12Ni6C036 | 13.16 | 6.69 | 100% Martensite |
| Cr12Ni8C036 | 13.54 | 9.16 | 90% Martensite +10% Austenite |

## 3. Tested method

For testing the cavitation erosion resistance there were used two vibratory devices, the first T1, of magnetostrictive type with nickel tube (vibration amplitude 94 μm, vibration frequency 7000 Hz, specimen diameter 14 mm, power of the electronic generator 500 W) and the second T2, constructed by taking into account the recommendations of ASTM G32-2010 Standard (vibration amplitude 50 μm, vibration frequency 20,000 Hz, specimen diameter 15.8 mm, power of the electronic generator 500 W). Both devices were realized at the Hydraulic Machinery Laboratory of the Timisoara Polytechic University [3].

The testing liquid was drinking water closer to that use by the field hydraulic turbines and pumps.

Total cavitation exposure was 2.75 hours divided in intermediary periods (one of 0.083 h, one of 0.167 h and ten of 0.25 h).

The testing procedure respect the ASTM G32-2010 Standard [8].

## 4. Experimental results and discussions

Figure 1 and 2 present the correlations between the cavitation erosion resistance 1/MDER and the principal mechanical properties of the eight tested steels, for both testing facilities.

From the Figures 1 and 2 results that for some steels the dependence of cavitation erosion resistance evolve with the mechanical characteristic values approximate in the form of power functions. This behavior is in accordance with those obtained by Garcia and Hammitt [5]. The steels with 0.036 %C present small scatters from the curve The best cavitation resistance was obtained for Cr12Ni8 with 10% austenite and the worst for Cr12Ni2 with 45% ferrite. These results confirm the conclusion that ferrite is the weakest structure to cavitation erosion. For the steels with 0.1 % C, the brand Cr12Ni6 (structure 40% M+60%A) present the outmost resistance to cavitation erosion and excellent mechanical properties. The rest of the steels present great scatters from curve. Cr 12Ni8 (structure 100% A) has reduced mechanical characteristics but very good cavitation erosion resistance. The steel Cr 12Ni05 (25% F) has good mechanical characteristics but poor cavitation resistance. The important scatter of the results for the steels with 0.1 % of carbon shows that cavitation erosion tests are imperious necessary.

IOP Conf. Series: Materials Science and Engineering **57** (2014) 012006 doi:10.1088/1757-899X/57/1/012006

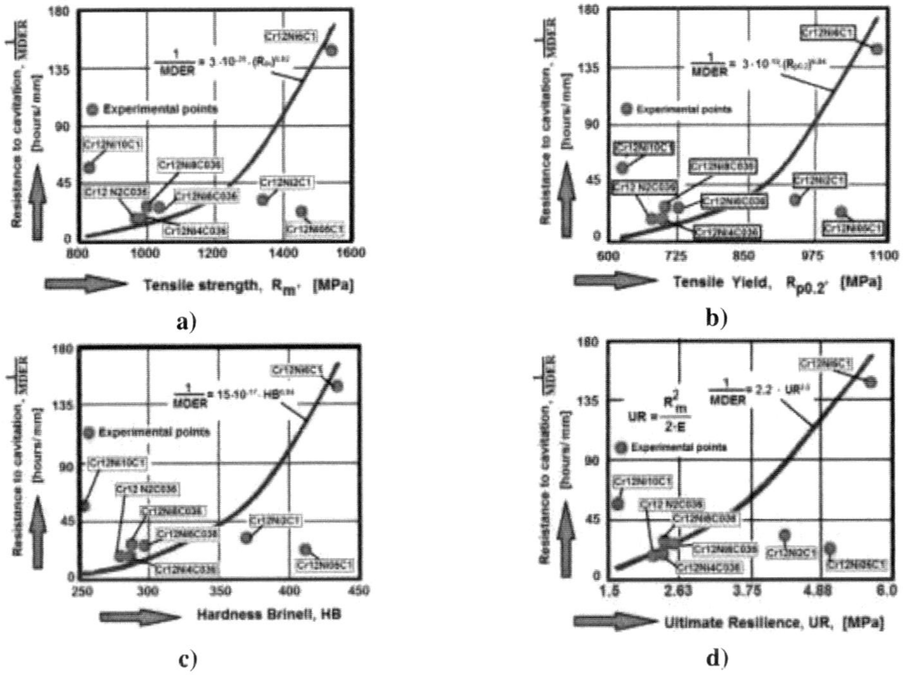


**Figure 1.** Correlation between cavitation erosion resistance and the principal mechanical properties (device T1); – tensile strength, b) - yield point, c) - Brinell hardness, d) – ultimate resilience

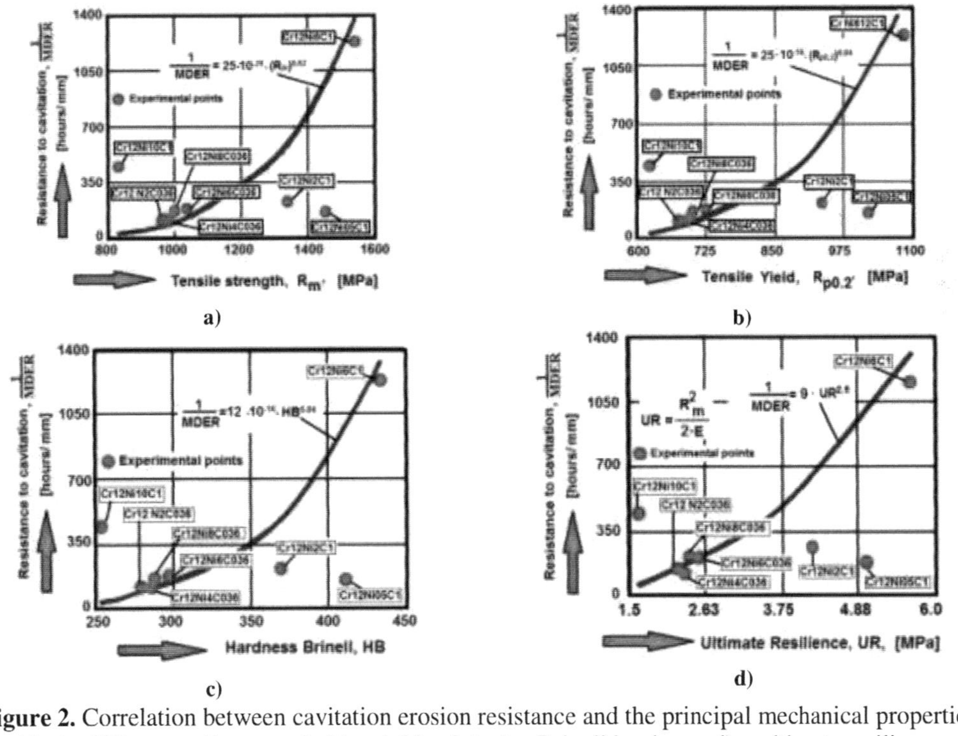


**Figure 2.** Correlation between cavitation erosion resistance and the principal mechanical properties (device T2); – tensile strength, b) - yield point, c) - Brinell hardness, d) – ultimate resilience

The principal cause of the scatter is the proportion of unstable austenite which at the stresses created by the bubble implosions is transformed suddenly in martensite exactly in the zone were the erosion take place and determines a local increase at cavitation erosion. Because the steels with 0.1% C show good cavitation erosion resistance, in the future for the steels used for important cavitation attack, the reduction of carbon content must not go to great values under 0.1% C.
The results obtained in the present work show that regardless of the vibratory device used, T1 or T2, there are no great differences in comparing various steels, even if both value of the MDE and MDER present great differences (approximately 1/8).

## 5. Conclusions

1. The increase of the values for principal mechanical characteristics ($R_m$, $R_{p0.2}$, hardness), in general, reduce the cavitation erosion.
2. The correlations between the parameter 1/MDER and the mechanical characteristics are similar with those established by Garcia and Hammitt, but the values the exponents depend on the tested material.
3. The values of the scale and shape parameters of the curves "mechanical properties against erosion" varies with the cavitation intensity of the test facility.
4. For simultaneously good cavitation erosion and an easy welding, the steels used for the hydraulic turbines blades or pump impellers must have the carbon content under 0.1 but in immediate vicinity of this value.
5. The scatter of the results for the steels with 0.1 % C, regardless of the parameters taken into account ($R_m$, $R_{p0.2}$, etc) show that such tests remain necessary also in the future, because for some stainless steels the increase of mechanical parameters are not proportional with the increase of cavitation erosion.
6. The principal cause of the scatter observed is the unstable austenite which at the stresses created by the bubble implosions is transformed suddenly in martensite and determines a local increase at cavitation erosion.

## References

[1] Anton I 1984 *Cavitatia* Vol I (Bucuresti: Academiei RSR)
[2] Anton I 1985 *Cavitatia* Vol II (Bucuresti: Academiei RSR)
[3] Bordeaşu I 2006 *Eroziunea cavitaţională a materialelor* (Timisoara: Politehnica)
[4] Franc J P 1995 *La Cavitation. Mecanismes phisiques et aspects industriels* (Grenoble: Press Universitaires de Grenoble)
[5] Garcia R, Hammitt F G and Nystrom R E 1960 *Correlation of cavitation damage with other material and fluid properties, Erosion by Cavitation or Impingement* (Atlantic City: ASTM, STP 408)
[6] Jurchela A 2012 *Cercetări asupa eroziunii produsă prin cavitaţie vibratorie la oţelurile inoxidabile cu conţinut constant în crom şi variabil în nichel* (Timisoara: PhD Thesis, Politehnica University of Timisoara)
[7] Mitelea I 1983 *Studiul metalelor* (Timisoara: Litografia Institutului Politehnic"Traian Vuia")
[8] *** 2010 Standard Test Method for Cavitation Erosion Using Vibratory Apparatus ASTM G3

# Cavitation erosion resistance of two steels with the same percentage of Chromium and Nickel but different Carbon content

**I Bordeasu[1], M O Popoviciu[2], I Mitelea[1], B Ghiban[3] and N Ghiban[3]**

[1] „Politehnica“ University of Timisoara, Bvd. Mihai Viteazul No.1, 300222 Timisoara, Romania
[2] Academy of Romanian Scientists, Bvd. Mihai Viteazul No.1, 300222 Timisoara, Romania
[3] Politehnica University of Bucharest, Splaiul Independentei No.313, 060042 Bucharest, Romania,

E-mail: ilarica59@gmail.com

**Abstract**. Hydraulic machinery repair works necessitate new materials with improved cavitation erosion resistance and simultaneously good welding properties. The present paper is concerned with the behavior at cavitation of two steels with close contents of chromium (approximately 12%) and nickel (approximate 6%) but with different carbon content (for the first 0.1% C and the second 0.036% C). The reduced carbon content is necessary for an easy welding repair work. As a result of the different chemical content, as is shown by the Schäffler diagram, the steel containing 0.1% C has a structure formed by 60% austenite and 40% martensite while those with 0.036% C has completely martensitic structure. The laboratory tests were done in two vibratory devices one with piezoelectric crystals, respecting the ASME G32-2010 Standard and the other a magnetostrictive vibratory device with nickel tube. The evaluation of the cavitation resistance was obtained with the help of cavitation erosion characteristic curves MDE (t) and MDER (t). For analyzing the steel degradation, the eroded areas were also subjected to microscopic investigations. The results show that the steel with 0.1% C has better cavitation erosion behavior than that of the steel with 0.036% C.

## 1. Introduction

Two great problems must be solved by the researchers, the manufacturers and the maintenance personnel, namely the time increase of the good running of the machinery and the reduction of the repair working time. This implies the use of steels with excellent cavitation erosion qualities but in the same time with the capacity to accept easy welding of the eroded zones. From the cavitation erosion resistance, the researches and the field experience show that the stainless steels with martensitic structure are suited for this purpose both from the point of view of resistance and costs [1], [3]. The welding repair work is not a very easy one but is possible especially by using electrodes with austenite or austenite ferrite structure, the electrodes cost is high but the quantities employed is not very great. Another problem is the carbon content of the base material. In most cases this content is chosen under 0.1% in order to reduce the welding difficulties and to avoid post welding cracks. This was the base to subject to laboratory tests two steels with different carbon contents (0.112% and 0.36%) but the same chromium (12%) and nickel (6%) contents. Test results obtained with two vibratory devices, one having a nickel magnetostrictive tube (noted T1) [2] and the other with piezoelectric crystals respecting the recent ASTM G-32 2010 Standards [9] put into evidence the influence of carbon content. It resulted that the carbon content must not be extremely reduced because it severely diminishes the cavitation erosion qualities of the steel.

## 2. Tested materials and testing facility. Discussion

The tested specimens were manufactured from cast bars after a mixing formula conceived by the experts from the Bucharest Polytechnic University, Center of Special Materials Survey. In the following, for these stainless steels were use the notations:

**1** - Cr12Ni6C1- for steels with **12,06 % chromium, 5.95 % nickel, 0.112 % carbon**, 1.67 % manganese, 0.023 % cobalt, 1.69 % silicon, 0.031 % molybdenum, 0.01 % tantalum , 0.031 % niobium, 0.047 % titan, 0.047 % vanadium, 0.016 % wolfram, 0.92 % aluminum, about 76.94 % iron, the rest being small quantities of accompanying chemical elements.

**2** - Cr12Ni6C036 - for steels with **12.059 % chromium, 5.597 % nickel, 0.036 % carbon**, 0.28 % manganese, 0.002 % cobalt, 0.461 % silicon, 0.039 % molybdenum, 0.009 % niobium, 0.073 % titan, 0.009 % vanadium, 0.153 % wolfram, 0.064% aluminum, about 81.37 % iron, the rest being small quantities of accompanying chemical elements.

The manufacturing processes of the specimens inclusive the heat treatments were identical for both materials [2], [6], [7], [8]. The chemical composition and the resulted structure give differences between the mechanical properties [4], which are not analyzed in the present paper. The testing liquid was water taken from the urban grid at 22±1°C. The resistance to cavitation erosion was obtained in two vibratory devices with different parameters. The T1 device is of the magnetostrictive type with nickel tube (vibration amplitude 94 μm, frequency 7000 Hz, specimen diameter 14 mm, power of the ultrasonic generator 500 W). The T2 device was recently realized, respecting the ASTM G32-2010 Standard, has piezoelectric crystals (vibration amplitude 50 μm, frequency 20 kHz, specimen diameter 15.8 mm, power of the ultrasonic generator 500 W) [5], [9]. Both test facilities were realized in the Cavitation Laboratory of Timisoara Polytechnic University.

The testing procedure (cleaning, washing, drying, weighing and specimen maintenance) and the duration to cavitation exposure (total exposure 2.75 hours with two intermediary periods of 0.83, 0.167 and the rest of 0.25 hours) were those used in the last 40 years in our laboratory.

The experimental results are presented by the characteristic curves (Fig. 1 and 2) which gives the time evolutions of both cumulative mean depth erosions (MDE) and mean depth erosion rates (MDER) for the tested materials. Supplementary are given histograms which present comparisons between the mean depth cavitation resistance 1/MDER of the tested steels (Fig.3) as well as comparisons between mean depth erosions (MDE) and maximum depth erosions MxDE). This comparisons are presented for both testing facilities (Fig. 4). The MxDE were microscopically measured in axial cross sections carefully selected.

The parameters MDE and MDER were computed from the mass loss recorded after each testing period by using the following relations:

$$\text{MDE} = \sum_{i=1}^{12} \left(\frac{4 \cdot \Delta M_i}{\rho \cdot \pi \cdot d_p^2}\right) \text{ [mm]}$$

$$\text{MDER} = \frac{4 \cdot \Delta M_i \cdot 60}{\rho \cdot \pi \cdot d_p^2 \cdot \Delta t_i} \text{ [mm/hours]} \qquad (1)$$

where:

$\Delta M_i$ - is the mass loss of the period “i”, measured in grams,

ρ – is the steel density in grams/mm$^3$,

$\Delta t_i$ – is the the duration of the period “i”,

i = 1,2, 3...12 – is the cavitation period (for i = 1, $\Delta t$ = 0.083 hours, for i = 2, $\Delta t$= 0.167 h, and for i= 3...12, $\Delta t$ = 0.250 h is equal for each period),

$d_p$ – is the specimen diameter in mm.

The evolutions of the curves MDE(t) and MDER (t) show the behavior of the used steels at mechanical stresses and local fatigue, generated by the impact with the forces developed through the bubble implosions.

In order to establish the performances of the two tested steels in Fig. 1 and 2 is given also the behavior of the steel OH12NDL, with martensitic structure, used in the past for manufacturing numerous hydraulic turbines in our country [1], [2].

The parameter 1/MDER chosen to characterize the steels resistance at cavitation erosion is in most cases the inverse of the stabilization MDER value and in some rare cases the MDER value for the maximum exposure time.

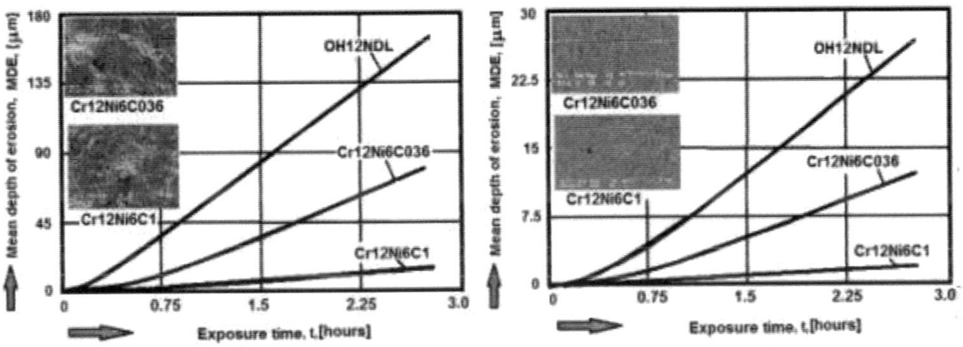


Test facility T1 Test facility T2

**Figure 1** Mean depth erosion against exposure time

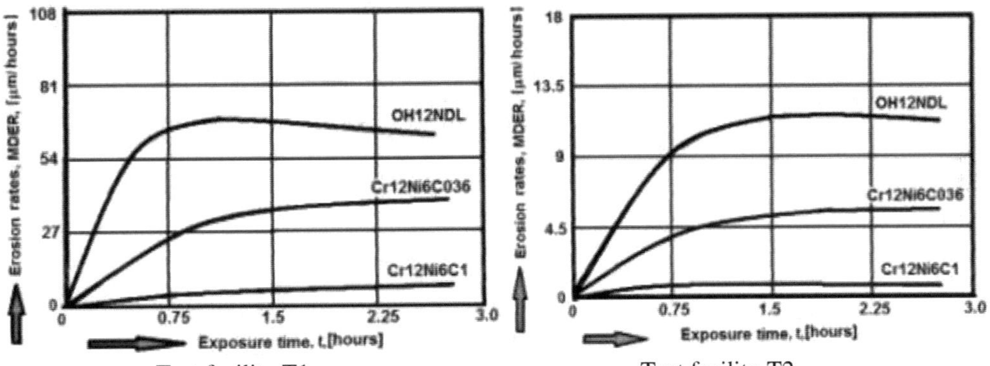


Test facility T1 Test facility T2

**Figure 2** Mean depth erosion rate against exposure time

IOP Conf. Series: Materials Science and Engineering **57** (2014) 012007 doi:10.1088/1757-899X/57/1/012007

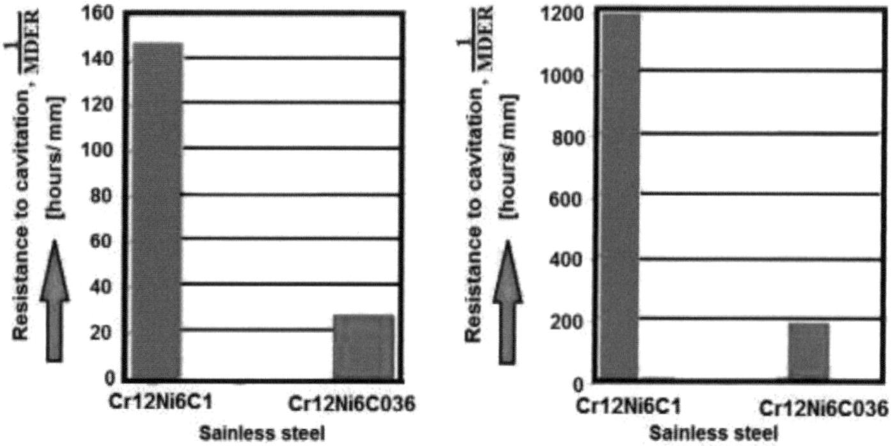


(Test facility T1) (Test facility T2)

**Figure 3** Comparisons of cavitation erosion resistances

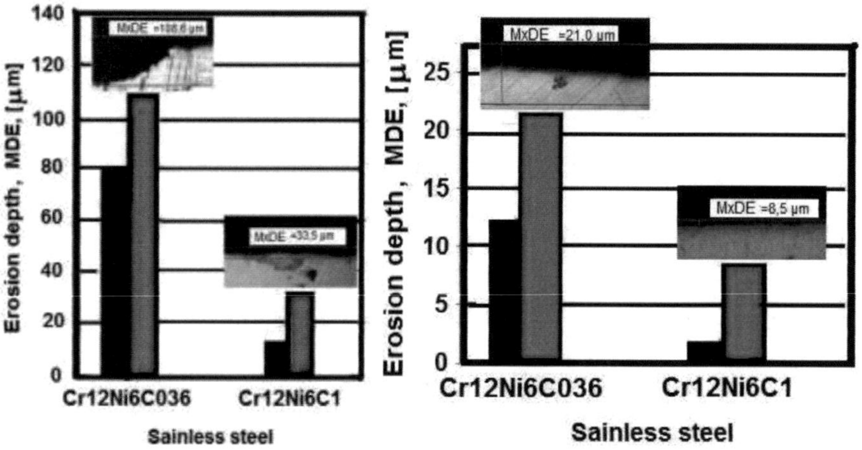


Test facility T1 (Test facility T2)

**Figure 4** Comparisons between the erosions depth
1- computed (MDE), 2- measured (MxDE)

From the diagrams it result the following conclusions:

- Regardless of used facility for generating vibratory cavitation (T1 or T2) the time evolution of the curves MDE(t) Fig. 1, and MDER(t) Fig. 2 is similar. The differences between the erosion mean depth erosion values is given by the difference of the parameters devices. The erosion intensity of T1 is greater than those of T2 and that gives for T1 greater cavern depth, especially the maximum depth is impressive. On the other hand, the eroded the area resulted with T2 covers approximate 85% of the whole area subjected to cavitation while for T1 only 60% of the area is affected.

- The mean depth erosion rates have the tendency to become stable for the maximum attained value. This behavior, in conformity with the former researches undertaken in Cavitation Laboratory of Timisoara Polytechnic University [1], [2], [5], is specific for steels with high cavitation erosion resistance, or for those with hardened superficial layer.

- By comparing the cavitation erosion resistances Fig. 3 or the maximum values for the mean depth erosions, Fig.2 it has been ascertained that the smallest cavitation erosion resistance is that of the steel

Cr12Ni16C036. The explanation is the low content of carbon which reduces the hardness of the area exposed to cavitation [5].

- By comparing the cavitation erosion behavior of the tested steel with that of OH12NDL steel resulted that both have better resistance and can be recommended for manufacturing of hydraulic turbines runners and blades.

- Although, in literature [1], [3], [4] the martensite steels are considered to have the highest cavitation erosion behavior, the data presented in Fig. 3 and 4 show that some steels with mix structure such as Cr12Ni6C1 with 60% austenite and 40% martensite exhibit superior qualities from this point of view and can replace the Cr12Ni6C036 steels used long time for manufacturing pump impellers or turbine runners [1], [2]. Because austenite has excellent welding capabilities, in the future researches, steels with different structure as the martensitic one must be analyzed, taking into account also the manufacturing qualities, mechanical characteristics, and the prime cost. The carbon content must be under 0.1% but not very low, because it gives also good cavitation erosion qualities.

- The comparisons between the mean depth erosion (MDE) and the maximum depth erosion (MxDE) must be maintained in use. The value of MDE is the important number for comparisons, but for two steels with the same MDE but different MxDE the better is those with low values of maximum depth of erosion.

To have an image upon the effect of the test facilities parameters upon the resulted cavitation erosion in Table 1 there are presented the ratios between the mean depth erosions rates (in stable zone) for the researched steels.

**Table 1** Mean depth erosion rates ratio

| **Ratio** | **Steel** | |
|---|---|---|
| | **Cr12Ni6C1** | **Cr12Ni6C036** |
| $\frac{MDERs_{T1}}{MDERs_{T2}}$ | 8,071 | 6,163 |

From table 1 it can be seen that the cavitations erosion intensity is greater for the device with nickel tube. Examining with a scanning electronic microscope the eroded areas of the Cr12Ni6C1 steel specimens both for T1 and T2 (see Fig.1) there were observed uniform erosions and zones with fragile rupture aspect. This is the result of the fact that both martensite and austentite with small quantities of chromium carbides in the structure assure a great cavitation erosion resistance. Also, the eroded area has a mix aspect with fine caverns (1…5 μm), cleavages surfaces and inter granular cracks. The same examinations made for the specimens from Cr12Ni6C036, regardless if they are eroded with the facility T1 or T2 (see Fig. 1) show erosions with irregular destructions and big caverns.

## 3. Conclusions

1. Laboratory results show that both tested steels present very good cavitation erosion resistance and can be used for manufacturing pieces heavy exposed to such strains as for example hydraulic machinery runners and blades.
2. The tests effected on two different facilities T1 and T2 show that Cr12Ni6C1 steel with an initial 60% austenite and 40% martensite structure has better cavitation erosion resistance than Cr12Ni6C1 steel with an initial 100% martensite structure.
3. In the study were not considered the changes that occur in the steels structure due to mechanical impact of the attack surface with shock waves or micro jets generated by the implosions of cavitation bubbles.
4. Regardless of the different running parameters, both test facilities, give approximate the same result when the used stainless steels are compared.

**References**

[1] Anton I 1985 *Cavitatia* Vol II (Bucuresti: Editura Academiei RSR)
[2] Bordeasu I 2006 *Eroziunea cavitațională a materialelor* (Timisoara: Editura Politehnica)
[3] Franc J P 1995 *La Cavitation, Mécanismes physiques et aspects industriels* (Grenoble: Presse Universitaires de Grenoble)
[4] Garcia R 1996 *Comprehensive Cavitation damage Data for Water and Various Liquid Metals Including Correlation with Material and Fluid Properties* (Michigan: Technical Rapport nr.6, University of Michigan)
[5] Jurchela A D, Bordeasu I, Mitelea I and Karabenciov A 2012 Considerations on the Effects of Carbon Content on the Cavitation Erosion Resistance of Stainless Steels with Controled Content of Chromium and Carbon *Proceedings of 21st International Conference on Metallurgy and Materials (Brno, Czech Republic, 23-25 May 2012)* pp 718-724
[6] Mitelea I, Bordeasu I, Popoviciu M and Hadar A 2007 Corrosion of Stainless Steels with "Soft" martensitic Structure *Revista de chimie* **58** 254
[7] Mitelea I, 1983 *Studiul metalelor* (Timisoara: Litografia Institutului Politehnic" Traian Vuia")
[8] Simoneau R 1987 Cavitation Erosion and Deformation mechanisms of Ni and Co Austenitic Stainless Steels, Erosion by Liquid and Solid Impact *Proceedings of ELSI-VII (Cambridge, UK, 7-10 October 1987)* pp 32.1-32.8
[9] *** 2010 Standard Test Method for Cavitation Erosion Using Vibratory Apparatus ASTM G-32

IOP Conf. Series: Materials Science and Engineering 57 (2014) 012008 doi:10.1088/1757-899X/57/1/012008

# Improved FOCUSS method with two-step reweighted $\ell_2$ minimization

**X Han[1] and Z Jiang[1]**

[1] Air Force Airborne Academy, Guilin, P.R.China

E-mail: thuhxb@gmail.com

**Abstract.** FOCal Underdetermined System Solver (FOCUSS) is a useful method through reweighted $\ell_2$ minimization for sparse recovery. In this paper, we introduce an improved FOCUSS by enhancing sparsity with two reweighted $\ell_2$ minimization. The reweighted FOCUSS method has higher mission success rate and better accuracy than FOCUSS. The simulation results illustrate the advantage of reweighted FOCUSS.

## 1. Introduction

The problems of finding sparse solutions to underdetermined system have been the hot spot because of the increasing applications in many fields, such as signal processing, image processing, information encoding, etc. This problem can be modeled as:

$$\mathbf{y} = \mathbf{A}\mathbf{x} + \mathbf{n} \tag{1}$$

where $\mathbf{y} = (y_1, \cdots, y_m)^T \to \mathbf{R}^m$ is an observable vector, $\mathbf{x} = (x_1, \ldots, x_n)^T \to \mathbf{R}^n$ is an unknown input vector, $\mathbf{A}$ is a known $m \times n$ basis matrix, and $\mathbf{n}$ represents noise vector.

The main objective is to recover $\mathbf{x}$ from (1) such that $\mathbf{x}$ is sparse, $m \times n$, and $\mathbf{A}$ satisfies some property (e.g. RIP [1]).

In the noise-free setup, it has been asserted that exact reconstruction of $\mathbf{x}$ can be guaranteed by solving a convex problem [1]:

$$\min_x \|\mathbf{x}\|_1 \text{ s.t. } \mathbf{y}=\mathbf{A}\mathbf{x}.$$

For (1), the solution of approximate reconstruction can be expressed by another convex problem with Lagrange form as

$$\min_x \|y - Ax\|_{\ell_2}^2 + \lambda\|x\|_{\ell_1}.$$

However, the computational burden of convex optimization is very high, and thus unsuitable to large scale problems and practice applications. At the same time, many substitutive algorithms have been proposed for this problem, such as greedy algorithms (e.g., MP, OMP [2], CoSaMP [3], etc.), FOCUSS algorithm(s) [4]. The performance of greedy algorithms can only be guaranteed only if $\mathbf{A}$ satisfies rigorous RIP condition and usually a little worse than others; FOCUSS algorithms are advantageous in terms of computational complexity and loose requirement on $\mathbf{A}$.

In the correspondence, FOCUSS algorithms are worth improving. As mentioned that FOCUSS is an iterative procedure of reweighted $\ell_2$ minimization, the main contribution of this paper is to improve FOCUSS with another iterative reweighted procedure.

## 2. Drawback analysis of FOCUSS

**Table 1.** FOCUSS algorithms

| basic FOCUSS | regularized FOCUSS |
|---|---|
| step 1: $\mathbf{W}_k = \mathrm{diag}(\mathbf{x}_{k-1})$<br>step 2: $\mathbf{q}_k = (\mathbf{AW}_k)^{\dagger}\mathbf{y}$,<br>step 3: $\mathbf{x}_k = \mathbf{W}_k\mathbf{q}_k$ | step 1: $\mathbf{W}_k = \mathrm{diag}(\lvert\mathbf{x}_{k-1}\rvert^{1-p/2})$<br>step 2: $\mathbf{q}_k = \mathbf{A}_k^H\left(\mathbf{A}_k\mathbf{A}_k^H + \lambda\mathbf{I}\right)^{-1}\mathbf{y}$,<br>*where* $\mathbf{A}_k = \mathbf{AW}_k$<br>step 3: $\mathbf{x}_k = \mathbf{W}_k\mathbf{q}_k$ |

The basic form and regularized form of FOCUSS algorithm are described [4] as Table 1. In Table 1, $\lambda$ is Lagrange parameter [5] and $0 < p \le 1$ is pre-set norm-factor. FOCUSS algorithm(s) can be considered a reweighted $\ell_2$ minimization because in each iteration, it is equivalent to find the optimum solutions to (2) [4,5].

$$\min\|\mathbf{q}_k\|_{\ell_2}^2, \text{ s.t. } \|\mathbf{Ax}-\mathbf{y}\|_{\ell_2} \le \beta$$
$$\text{where } \|\mathbf{q}_k\|_{\ell_2}^2 = \|\mathbf{Wx}_k\|_{\ell_2}^2 = \sum_{i=1}^{n}\left|\frac{x_{k,i}}{x_{k-1,i}}\right|^2 \tag{2}$$

where $\beta$ is noise threshold, and $\mathbf{W} = \mathbf{W}_k^{-1}$.

The reweighted method of FOCUSS is accordant to idea of enhancing sparsity by reweighted $\ell_1$ minimization [6], which mentions that large entries in $w_i$ force the solution $\mathbf{x}$ to concentrate on the indices where $w_i$ is small in order to approach $\ell_0$ minimization further more.

$\mathbf{x}_k$ in Table 1 is equivalent to

$$\mathbf{x}_k = \mathbf{W}_k\left(\mathbf{W}_k^H\mathbf{A}^H\mathbf{A}W_k\right)^{-1}\mathbf{W}_k^H\mathbf{A}^H\mathbf{y}$$
$$\text{or } \mathbf{x}_k = \mathbf{W}_k\left(\mathbf{W}_k^H\mathbf{A}^H\mathbf{A}W_k + \lambda\mathbf{I}\right)^{-1}\mathbf{W}_k^H\mathbf{A}^H\mathbf{y} \tag{3}$$

From (3) we can find that $\mathbf{x}_k$, the approximate solution of $\mathbf{x}$ in $k$-th iteration depends on the constant entries $\mathbf{A}^H\mathbf{y}$. Let $\mathbf{A} = [\mathbf{a}_1, \cdots, \mathbf{a}_n]$, $\mathbf{T}$ represents the indices set of non-zero entries, $\bar{\mathbf{T}}$ represents the indices set of zero entries. Suppose $t \in \mathbf{T}$ is corresponding to the index of the smallest non-zero entry in $\mathbf{x}$, $\hat{x}_{k,t}$, the estimation of $x_t$ is desponded on $\mathbf{a}_t^H\mathbf{y}$. Due to

$$\mathbf{a}_t^H\mathbf{y} = \mathbf{a}_t^H\mathbf{Ax} + \mathbf{a}_t^H\mathbf{n}, \tag{4}$$

if $\mathbf{a}_t^H\mathbf{Ax}$ and $\mathbf{a}_t^H\mathbf{n}$ have the opposite phase, the power of signal in $\mathbf{a}_t^H\mathbf{y}$ would be crippled even counteracted. So with some possibility, there exists $d \in \bar{\mathbf{T}}$ satisfying $|\hat{x}_{k,t}| < |\hat{x}_{k,d}|$. Under this situation, $w_t$, which should be smaller than $w_d$, is bigger than $w_d$. This situation will never be improved through iterations, and thus leads to the failure of FOCUSS.

## 3. Reweighted FOCUSS algorithm

In order to overcome the drawback in Section 2, we design the reweighted FOCUSS which utilize another different reweighted $\ell_2$ minimization. Similar to [6] and [7], the important thing of reweighted FOCUSS is assigning bigger weights to those elements of $\mathbf{x}$ which are more likely to be zero. Reweighted FOCUSS algorithm is described in Table 2.

In reweighted FOCUSS algorithm, FOCUSS algorithms can be considered the first step of reweighted $\ell_2$ minimization and the following procedures can be considered the second step which is equivalent to find the optimum solutions to the following:

**Table 2**. Reweighted FOCUSS algorithm

**Algorithm** *reweighted FOCUSS:*

1. Get the initial estimation of $\mathbf{x}_0$ of $\mathbf{x}$ through FOCUSS algorithms;
2. Find the index set $\mathbf{K} \subset \{1, 2, \cdots, n\}$ which corresponds to the largest $C \cdot \frac{m}{\log(n/m)}$ elements of $\mathbf{x}_k$ in amplitudes [8], where $C$ is a constant with the numerical value given in [1];
3. Set diagonal matrix $\Omega$ with diagonal elements satisfying

$$\begin{cases} \omega_i - 1, i \in \mathbf{K} \\ \omega_i > 1, i \in \bar{\mathbf{K}} \end{cases}$$

and calculate

$$\mathbf{p}_k = (\mathbf{A}\Omega)^{\dagger} \mathbf{y} \ \text{ or } \mathbf{p}_k = \Omega^H \Omega^H \left(\mathbf{A}\Omega\Omega^H \mathbf{A}^H + \lambda \mathbf{I}\right)^{-1} \mathbf{y},$$
$$\mathbf{x}_k = \Omega \mathbf{p}_k;$$

4. Terminate on convergence or when $k$ attains a specified maximum number of iterations $k_{\max}$; otherwise, increase $k$ and go to step 2.

$$\min \|\mathbf{p}_k\|_{\ell_2}^2, \text{ s.t. } \|\mathbf{Ax} - \mathbf{y}\|_{\ell_2} \leq \beta$$
$$\text{where } \|\mathbf{p}_k\|_{\ell_2}^2 = \|\mathbf{Wx}_k\|_{\ell_2}^2 = \sum_{i \in \mathbf{K}} |x_{k,i}|^2 + \sum_{i \in \bar{\mathbf{K}}} |\omega_i x_{k,i}|^2 \tag{5}$$

The method choosing $\mathbf{W}$ in (5) is proved to be efficient to recover $\mathbf{x}$ from (1) in [1]. However, differently from [1] which needs solving convex problem, this paper employs iterated method to solve the reweighted $\ell_2$ minimization approximately, which deduces the computational burden greatly.

Through re-assigning the weighted coefficient, if some indices of small nonzero entries in $\mathbf{x}$ are undetected or some indices of zero entries are false-drop under FOCUSS algorithms just like the analysis in Section 2, they can be corrected.

## 4. Simulation results

This section shows the advantages of recovery ability of reweighted FOCUSS algorithms with numerical simulations, compared to FOCUSS algorithms. In a Monte Carlo simulation, 1000 trials are carried out independently. In each trial, $\mathbf{A}$ is chosen as Gaussian random matrix, entries of which are independently, identically and normally distributed; $\mathbf{x}$ is a $s$-sparse vector which is input randomly, with normalizing power, i.e., $\sum_i |x_i|^2 = 1$; entries of $\mathbf{n}$ are independently and identically Gaussian distributed with mean zero and variance $\sigma^2$; The indices of nonzero coordinate set $\mathbf{T}$ are chosen randomly from a discrete uniform distribution of [1,n]. Then SNR of system can be represented as $1/\sigma^2$. The algorithm in simulations is considered to be successful if all nonzero-locations of $\mathbf{x}$ are found exactly; otherwise the algorithm is considered to be failed.

The algorithms mentioned in the simulation are: FOCUSS, regularized FOCUSS, reweighted FOCUSS and reweighted regularized FOCUSS. Figure1 shows the statistical results of mission success rate, and Figure 2 shows the statistical root-mean-square error (RMSE) curve of signal amplitude recovery when algorithms can find the nonzero-coordinate $\mathbf{T}$ correctly under different SNR scenes. It can be seen from Figure 1 and Figure 2 that reweighted FOCUSS works more robustly and more accurately than FOCUSS, and so does reweighted regularized FOCUSS better than regularized FOCUSS. In Figure 1 and Figure 2, $m=32$, $n=128$, $s=3$, $p=0.5$.

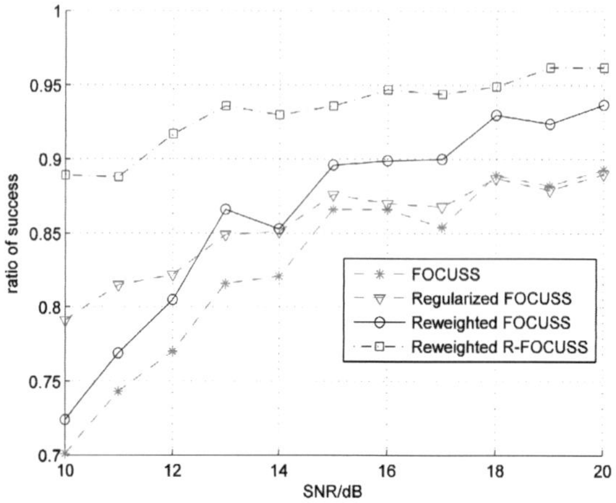


**Figure 1.** Mission success rate of algorithms in finding the support set correctly.

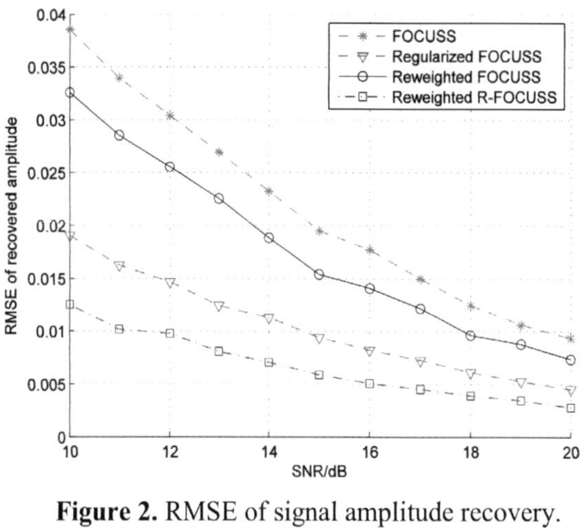


**Figure 2.** RMSE of signal amplitude recovery.

## 5. Conclusion

In this paper, for solving sparse-recovery problems, we have analyzed the major drawback of traditional FOCUSS algorithms, and developed the reweighted FOCUSS to approximate sparse solutions. We give the experimental evidence that reweighted FOCUSS can reconstruct the sparse inputs more robustly and accurately than untreated FOCUSS. This reweighted method has never conflict with sparse-recovery algorithms and can be regarded as the post-processing of them, and is easier to apply than reweighted $\ell_1$ minimization.

## References

[1] Candès E and Tao T 2005 Decoding by linear programming *IEEE Trans. Inf. Theory* **51** 4203
[2] Tropp J and Gilbert A 2007 *IEEE Trans. Inf. Theory* **53** 4655

[3] Needell D and Tropp J 2009 *Appl. Comput. Harmon. Anal.* **26** 301
[4] Gorodnitsky I and Rao B 1997 *IEEE Trans. Signal Process.* **45** 600
[5] Rao B, Engan K, Cotter S and Palmer J 2003 *IEEE Trans. Signal Process.* **51** 760
[6] Candès E, Wakin M and Boyd S 2008 *J. Fourier Anal Appl.* **14** 877
[7] Xu W, Khajehnejad M, Avestimehr A and Hassibi B 2010 *ICASSP* 5498
[8] Candès E, Romberg J and Tao T 2006 *Commun. Pure Appl. Math.* **59** 1207

IOP Conf. Series: Materials Science and Engineering **57** (2014) 012009 doi:10.1088/1757-899X/57/1/012009

# Dual rotor single- stator axial air gap PMSM motor/generator drive for high torque vehicles applications

**L N Tutelea[1], S I Deaconu[2], I Boldea[1] and G N Popa[2]**

[1] Politehnica" University of Timisoara, Electrical Engineering Department, Vasile Parvan str., no. 1-2, 300223 Timisoara, Romania

[2] Politehnica" University of Timisoara, Department of Electrotechnical Engineering and Industrial Informatics, 5 Revolution Street, Hunedoara, 331128, Romania

E-mail: sorin.deaconu@fih.upt.ro

**Abstract.** The actual e - continuously variable transmission (e-CVT) solution for the parallel Hybrid Electric Vehicle (HEV) requires two electric machines, two inverters, and a planetary gear. A distinct electric generator and a propulsion electric motor, both with full power converters, are typical for a series HEV. In an effort to simplify the planetary-geared e-CVT for the parallel HEV or the series HEV we hereby propose to replace the basically two electric machines and their two power converters by a single, axial-air-gap, electric machine central stator, fed from a single PWM converter with dual frequency voltage output and two independent PM rotors, destined for hybrid electric vehicles (HEV) and military vehicles applications. The proposed topologies and the magneto-motive force analysis are the core of the paper.

## 1. Introduction

The axial flux permanent magnet (AFPM) machine, also called the disc-type machine, is an attractive alternative due to its pancake shape, compact constructions and high power density. AFPM motors are particularly suitable for electrical vehicles, pumps, fans, valve control, centrifuges, machine tools, robots and industrial [9], [2].

Axial flux machines appeared in the technical literature in the early '70s and trading of axial flux induction motors started few years later [13], [4], [5], [11]. Nowadays, direct drive applications that require actuators or generators capable of operating at low speeds with large torques have revived the attention towards Axial Flux Machines, especially for the PM type, as they are capable of larger torque density and efficiency [13], [15], [8], [3], [6], [16].

However, AFPM Synchronous Machines become advantageous whenever a number of design prescriptions are fulfilled. Most notably, it is widely accepted that the number of pole pairs must be conveniently high [13], [12].

Fractional slot windings can be often realized in concentrated layouts: this happens when windings overhangs are not overlapped and the coils are wound individually around the stator teeth. Fractional Slot Concentrated Windings offer remarkable advantages both on the end user and to the manufacturer. In fact, they allow the physical separation of the phases and of the magnetic circuits of the phases, thus reducing the risk of phase-to-phase faults and minimizing the mutual inductance among the phases [13], [7].

The back iron cores of AFPM synchronous machines are usually made of laminated steel. However, Soft Magnetic Composites (SMC) is intensively studied since they are easy to machine and fit well higher frequencies [13].

The desired high-performance can be achieved by using PM machines and new control methods such as direct torque control (DTC) or other high performance vector control methods [14-15].

The present paper proposes a new dual PM rotor, SM drive with basically a single stator (with a Gramme-ring or dual winding, one on each side of stator) and dual rotor with different pole counts and high winding factors, to reduce volume, weight and cost in either planetary-gear parallel HEV or in series HEV.

## 2. Constructive elements

In figure 1 a 2D drawing of the machine in front view and a longitudinal section are shown.

The single stator dual PM rotors axial synchronous machine having in centre the stator assembly (1) with the two three-phase windings (2) placed in open slots, fixed rigid in the casing (3) provided with two side covers (4), (5) in which the two ball bearings supports (6), (7), one radial and one axial, are introduced. The ball bearings allow the two shafts (8), (9) to rotate independently, each shaft having in the side towards the stator a disk of solid steel on which the permanent magnet poles are placed in circular and symmetric manner . The other end of the shaft is inserted into a half-coupling which is connected to the thermal engine (10) respectively to the gears towards the drive wheels (11).

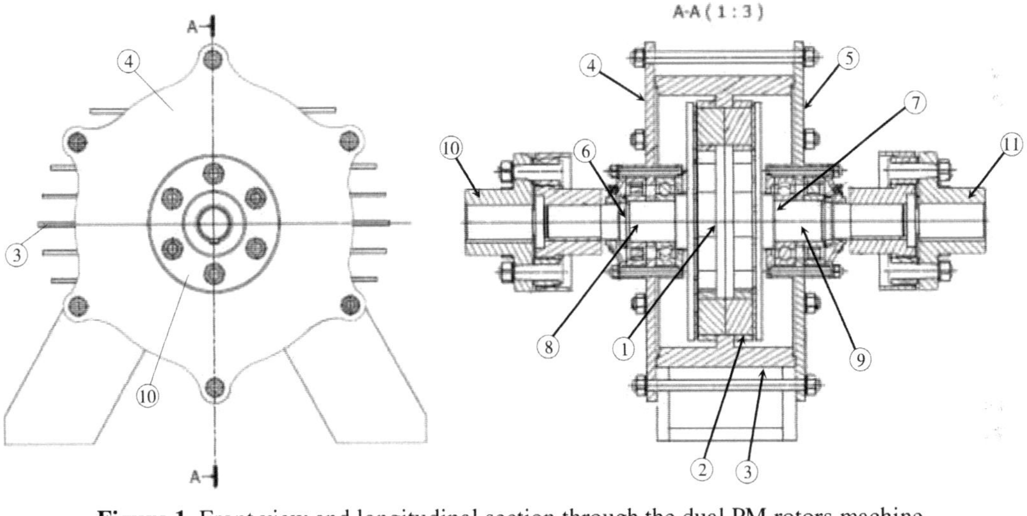


**Figure 1.** Front view and longitudinal section through the dual PM rotors machine.

For the stator magnetic core soft magnetic composite (SMC) is suggested, but a rolled-lamination magnetic core is also feasible. The same rationale is valid for the rotors magnetic core.

A surface-mounted PM topology has been considered for the rotors, 14 poles for rotor 1 and 10 poles for rotor 2 (figure 2).

The material of the permanent magnets is the Neodymium Iron Boron (Nd-Fe-B) type MPN 40SH, with flux density remanence $B_r$ = 1,13T and coercive field intensity $H_c = 860 \cdot 10^3 \, A/m$.

For this axial flux machine, the length of the air-gap g and the permanent magnet thickness $h_{PM}$ can be related to the other machine diameter $D_{out}$ by the expressions:

$$g = 6 \cdot 10^3 \cdot D_{out} \quad , \tag{1}$$

$$h_{PM} = \frac{4}{3} g \quad . \tag{2}$$

The machine designed in the paper has an external diameter $D_{out}$ of 260 mm and an air-gap equal to 2 mm. The thickness of the magnet is 5 mm.

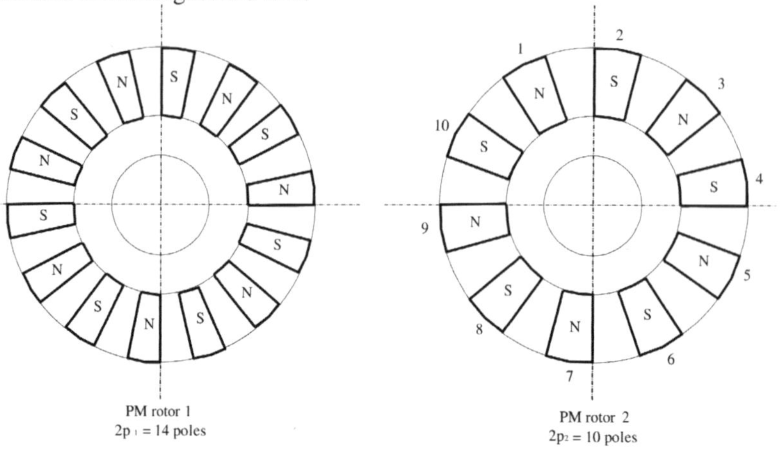


**Figure 2.** Rotor 1 with 14 PM poles (a) and rotor 2 with 10 PM poles (b).

## 3. Concentrated winding axial PM synchronous machine

### *3.1. Analytical analysis*

The distribution of the normal component of the magnetic flux density in the x-direction (circumferential) excited by PMs without armature reaction at the radius corresponding to the average pole pitch $\tau$ can be described by the following equation [9]:

$$B_g(x)=\frac{1}{k_c}\cdot B_{ZPM}(x)+B_{sl}(x)\ , \tag{3}$$

where the PM excitation flux density for smooth stator core is:

$$B_{ZPM}(x)=\sum_{v=1}^{\infty}\frac{1}{\sigma_{eM}}\frac{1}{k_c}B_g\cdot b_v\cdot\cos\left[v\left(\frac{\pi}{\tau}x-\frac{\pi}{2}\right)\right]\ , \tag{4}$$

$v = 1, 3, 5$, and the magnetic flux density component due to stator slots is:

$$B_{sl}(x)=\lambda_{sl}(x)\cdot\frac{g'}{\mu_0}B_{ZPM}(x)\ , \tag{5}$$

The peak values of the higher harmonics of the magnetic flux density distribution of equation (4) are $B_{mgv}=(B_g/\sigma_{eM})\cdot b_v$ [9].

In the equation (3), (4) and (5) $k_c$ is Carter's coefficient according to [9]:

$$k_c=\frac{t_1}{t_1-\gamma\cdot g}\ , \tag{6}$$

where $t_1$ is the average of slot pitch and $\gamma$ an geometrical coefficient, $b_v$ is according to [1]:

$$\begin{aligned}b_v=&\frac{4}{\tau}\left[\frac{c_p}{c_p^2+\beta_v^2}\sinh\alpha-6\left(\frac{1}{\beta_v}\right)^4\cdot\frac{1}{b_t^3}\cosh\alpha+3\left(\frac{1}{\beta_v}\right)^4\cdot\frac{2}{\tau-b_p}\cosh\alpha\right]\cdot\sin\left(v\cdot\frac{\pi}{2}\right)\cdot\\&\cdot\sin\left(v\cdot\frac{\pi b_t}{2\tau}\right)+\frac{4}{\tau}\left[\frac{\beta_v}{c_p^2+\beta_v^2}+6\left(\frac{1}{\beta_v}\right)^3\cdot\frac{1}{b_t^2}-\frac{1}{\beta_v}\right]\cdot\cosh\alpha\cdot\sin\left(v\cdot\frac{\pi}{2}\right)\cdot\cos\left(\beta_v\cdot\frac{b_t}{\tau}\right)\end{aligned} \tag{7}$$

$$\beta_v=v\cdot\frac{\pi}{2}\ , \tag{8}$$

$$b_t = \frac{\tau - b_p}{2}, \quad c_p = 2 \cdot \frac{\alpha}{b_p} \quad . \tag{9}$$

The parameter $\alpha$ depends on the shape of the distribution of the normal component of the magnetic flux density (usually $b_v$ is for $\alpha$=0). The parameter $\tau$ is the average pole pitch

$$\tau = \frac{\pi D}{2p} \quad , \tag{10}$$

where D is the average diameter

$$D = \frac{D_{out} + D_{in}}{2} \quad , \tag{11}$$

with $D_{in}$ the inner diameter of PMs.

With $k_{sat} \geq 1$ the saturation factor in the d axis and $k_{satq} \geq 1$ in the q axis and $\mu_{rrec}$ the relative recoil permeability of the PM [9]

$$g' = g \cdot k_c \cdot k_{sat} + \frac{h_{PM}}{\mu_{rrec}} \quad , \tag{12}$$

$$g_q' = g \cdot k_c \cdot k_{satq} + h_{PM} \quad , \tag{13}$$

$$B_g \approx \frac{B_r \cdot h_{PM}}{h_{PM} + \mu_{rrec} \cdot g} \quad . \tag{14}$$

$B_g$ is the flat-tapped value of the magnetic flux density, $\phi_M$ is the full equivalent flux of the PM and $\phi_g$ is the air gap magnetic flux:

$$\sigma_{eM} = \frac{\varphi_M}{\varphi_g} \quad , \tag{15}$$

and $\lambda_{sl}$ is the relative slot leakage permeance given by:

$$\lambda_{sl}(x) = -\sum_{v=1}^{\infty} a_v \cdot \cos\left( v \cdot \frac{2\pi}{t_1} \cdot x \right) \quad , \tag{16}$$

where

$$t_1 = \frac{2p \cdot \tau}{N_{st}} \quad , \tag{17}$$

and $a_v$ the amplitude of slot harmonics.

### *3.2. Armature windings design*

An AFPM machine can compete with the traditional radial-flux machines in terms of torque and power density if the number of pole-pairs is sufficiently high and the ratio axial length/outer diameter is low [9], [2], [16].

The term non-overlap (concentrated coil) windings defined here refer to the windings of which the coils do not overlap. As with overlap windings non-overlap windings can be of single or double layer, concentrated or distributed, integral or fractional and air-cored or iron-cored. In double layer, non-overlap slotted iron-cored windings two coils are sharing a slot (two coil layers per slot), which means that all teeth are wound. Consequently these windings are sometimes called tooth windings [9].

The various combinations of slots and poles which allow the realization of balanced winding can be determined by the relation:

$$\frac{N_{st}}{[GCD(N_{st}, 2p)]} = 3k \quad , \tag{18}$$

where $N_{st}$ is the number of slots, p the number of pair of poles and k an integer number. (GCD: Greatest Common Divisor). As is well known, the number of slots per pole and per phase of an electrical machine is equal to:

$$q = \frac{N_{st}}{2pm} , \tag{19}$$

where m is number of phases.

Fixing $N_{st}$, if the machine has a high number of pole pairs, q decreases.

Winding arrangements with a number of slots per pole and per phase lower than unity become sometimes mandatory for the construction of fractional power machines with large number of pole pairs. In some cases, adopting a layout of the kind makes it possible to realize concentrated non overlapping windings that at the same time yield high values of the fundamental winding factor.

In concentrated non-overlap windings, there is only one coil in a coil group or coil phase belt (z=1). The distribution factor for the fundamental space harmonic is calculated by [9]:

$$k_{d1} = \frac{\sin\left(\frac{\pi}{2m}\right)}{z \cdot \sin\left(\frac{\pi}{2mz}\right)} , \tag{20}$$

and the pitch factor for the fundamental space harmonic is [9]:

$$k_{p1} = \sin\left(\frac{\pi}{2} \cdot \frac{y}{\tau}\right) , \tag{21}$$

where y is the pitch of one coil and $\tau$ the pole pitch. The fundamental winding factor:

$$k_{w1} = k_{d1} \cdot k_{p1} . \tag{22}$$

For our case $k_{w1} = 0.966$ for both windings. ($N_{st}$=12, $p_1 = 7$, $p_2 = 5$).

The design of windings was performed by plotting the space vectors of the induced EMF in the slots and the windings arrangement. It is possible to envisage few simple rules to calculate the following parameters:

- the greatest common divider between the number of slots and the number of pole pairs

$$t' = GCD(N_{st}, p) = GCD(12,5) = 1 , \tag{23}$$

$$t'' = GCD(12,7) = 1 , \tag{24}$$

- the number of vectors of star

$$r' = \frac{N_{st}}{t} = \frac{12}{1} = 12 = r'' , \tag{25}$$

- the angle between two adjacent vectors

$$\alpha_1' = \frac{2\pi \cdot t'}{N_{st}} = \frac{\pi}{6} = \alpha_1'' , \tag{26}$$

- the phase lag between two vectors of two adjacent phases:

$$\alpha' = \frac{2\pi \cdot p}{N_{st}} = \frac{2\pi \cdot 5}{12} = \frac{5\pi}{6} , \tag{27}$$

$$\alpha'' = \frac{2\pi \cdot 7}{12} = \frac{7\pi}{6} . \tag{28}$$

### *3.3. MMF analysis*

The MMFs $F_{ad}$ and $F_{aq}$ in the d-axis and q-axis are [9]:

$$F_{ad} = \frac{m_1\sqrt{2}}{\pi} \cdot \frac{N_1 \cdot k_{w1}}{p} \cdot I_{ad} , \tag{29}$$

$$F_{aq} = \frac{m_1\sqrt{2}}{\pi} \cdot \frac{N_1 \cdot k_{w1}}{p} \cdot I_{aq} , \tag{30}$$

where $I_{ad}$ and $I_{aq}$ are the d and q axis stator (armature) currents respectively.

Let us consider the case where the concentrated winding PM machine is fed by sinusoidal currents, with a maximum current I in phase "A" ($i_A$=I and $i_B$=$i_C$=-I/2).

In order to investigate the spatial distribution of the armature MMF, the surface mounted PMs in the rotor are supposed to be discarded. Giving the fact that each phase is made up of N coils, the MMF $F_{A1}$ produced by one coil $\left(A^+, A^-\right)$ is expressed as [10]:

$$F_{A1}(\theta)=\begin{cases}(2-\beta)\cdot\dfrac{N\cdot I}{8}, & -\dfrac{\delta}{2}<\theta<\dfrac{\delta}{2}\\ -\beta\cdot\dfrac{N\cdot I}{8}, & \text{elsewhere}\end{cases}, \tag{31}$$

where $\beta$ is the coil pitch ratio, such that

$$\beta=\frac{\delta}{\tau_{st}}, \tag{32}$$

where $\tau_{st}$ is the armature pole pitch. Taking into account the hypothesis which considers that the magnetic circuit is not saturated, one can apply the superposition theorem in order to obtain the magneto-motive force (MMF) $F_A(\theta)$ of phase "A" as follows:

$$F_A(\theta)=F_{A1}(\theta)+F_{A2}(\theta)+F_{A3}(\theta)+F_{A4}(\theta), \tag{33}$$

With

$$\begin{cases}F_{A2}(\theta)=-F_{A1}\left(\theta-\dfrac{\pi}{6}\right)\\ F_{A3}(\theta)=-F_{A1}(\theta-\pi)\\ F_{A4}(\theta)=F_{A1}\left(\theta-\dfrac{7\pi}{6}\right)\end{cases}, \tag{34}$$

Now let us consider the spatial distribution of the MMFs produced by phases "B" and "C", witch are deduced from the one produced by phase "A" as follows:

$$\begin{cases}F_B(\theta)=-\dfrac{1}{2}F_A\left(\theta+\dfrac{2\pi}{3}\right)\\ F_C(\theta)=-\dfrac{1}{2}F_A\left(\theta-\dfrac{2\pi}{3}\right)\end{cases}. \tag{35}$$

## 4. Conclusions

Using an analytical design method or the 2D finite element analysis (FEA) only for the average radius on the machine does not generally yield sufficiently accurate computation results. With the 3D FEA, it is possible to take into consideration the actual 3D structure of the machine, but performing the computations is too often time-consuming, especially if the objective is to achieve a preliminary design of the machine. The basic idea of the hereby proposed design method is to firstly subdivide the axial flux machine into a sufficient amount of independent computation plans, next perform the required 2D computations on each plane, and finally compose the overall performance of the machine from the computation results obtained for each design plan.

The single stator dual-rotor permanent magnet axial synchronous machine can be controlled by a single inverter and two frequencies, the two rotors being able to operate both as motor and as generator in a wide speed range, in the same or opposite directions of motion. The transitory regime due to speed equalization in the two rotors is surpassed without catastrophic oscillations in torque and speed.

The peak value of the torque, the overload capability and very fast dynamic response make this electric drive system very good for military vehicles applications.

**References**

[1] Boldea I and Tutelea L N 2009 *Electric Machines. Steady State, Transients and Design with MATLAB* (Boca-Raton: CRC Press)

[2] Boldea I, Topor M, Marignetti F, Deaconu S I, and Tutelea L N 2010 A Novel, Single Stator Dual PM Rotor, Synchronous Machine: topology, circuit model, controlled dynamics simulation and 3D FEM Analysis of Torque Production *Proceedings of 12th International Conference on Optimization of Electrical and Electronic Equipment OPTIM (Brasov, Romania, 22-24 May 2010)* pp 343-351

[3] Brooking P and Bumby J R 2002 An integrated engine-generator set with power electronic interface for hybrid electric vehicle applications *Proceedings of International Conference on Power Electronics, Machines and Drives (Bath, UK, 2-7 June 2002)* pp 153-158

[4] Campbell P 1974 Principles of a permanent magnet axial-field DC machine *Proceedings of the Institution of Electrical Engineers* **121** 1489

[5] Campbell P 1975 The magnetic circuit of an axial field DC electrical machine *IEEE Transactions on Magnetics* **11** (5) 1541

[6] Cavagnino A, Lazzari M, Profumo F and Tenconi A 2002 A comparison between the axial flux and the radial flux structures for PM synchronous motors *IEEE Transaction on Industry Application* **38** (6) 1517

[7] Cros J and Viarouge P 2002 Synthesis of High performance PM Motors with Concentrated Windings *IEEE Transaction on Energy Conversion* **17** (2) 248

[8] Eastham J F, Profumo F, Tenconi A, Hill-Cottingham R J, Coles P C and Gianolio G 2002 Novel Axial flux Machine for aircraft drive: design and modeling *IEEE Transactions on Magnetics* **38** (5) 3003

[9] Gieras J F, Wang R J and Kamper M J 2008 *Axial Flux Permanent Brushless Machines* Second Edition (Berlin: Springer)

[10] Hamadou G B, Masmoudi A, Abdennadher I and Masmoudi A 2009 Design of a Single-Stator Dual-Rotor Permanent-Magnet Machine *IEEE Transactions on Magnetics* **45** 127

[11] Leung W 1978 Sandwich synchronous machines and DC machines *Patent specifications (London, UK, April 1978)*

[12] Marignetti F and Scarano M 1999 An Axial-flux PM Motor Wheel *Proceedings Electromotion '99 (Patras, Greece, 8-9 July 1999)* pp 1-6

[13] Marignetti F, Colli V D, Stefano R D and Cavagnino A 2007 Design Issues of a Fractional – Slot Windings Axial Flux PM Machine with Soft Magnetic Compound Stator *IECON 2007 (Taipei, Taiwan, 5-8 November 2007)* pp 187-192

[14] Parviainen A 2005 *Design of AFPM low-speed Machines and Performance Comparison between Radial-Flux and Axial-Flux Machines* (Lappeenranta: PhD Thesis, Lappeenranta University of Technology)

[15] Profumo F, Zhang Z and Tenconi A 1997 Axial flux machine drives a new viable solution for electric cars *IEEE Transaction on Industrial Electronics* **44** (1) 39

[16] Sitapati K and Krishnan R 2001 Performance Comparisons of Radial and Axial Field, permanent-Magnet, Brushless Machines *IEEE Transaction on Industry Application* **37** (5) 1219

# Generalized shrunken type-GM estimator and its application

**C Z Ma[1] and Y L Du[1]**

[1] Institute of Science, Information Engineering University, Zhengzhou 450001, China

E-mail: marcz@163.com

**Abstract.** The parameter estimation problem in linear model is considered when multicollinearity and outliers exist simultaneously. A class of new robust biased estimator, Generalized Shrunken Type-GM Estimation, with their calculated methods are established by combination of GM estimator and biased estimator include Ridge estimate, Principal components estimate and Liu estimate and so on. A numerical example shows that the most attractivc advantage of these new estimators is that they can not only overcome the multicollinearity of coefficient matrix and outliers but also have the ability to control the influence of leverage points.

## 1. Introduction

It is well known that the classical least squares (LS) estimation have many desirable statistical properties, particularly when error are normally distributed the variance of LS estimations are the least in all sort of unbiased estimation class. But LS estimator frequently are influenced by two class of problems that is outliers and multicollinearity , which in turn lead to highly unstable LS estimates. To circumvent influence of multicollinearity, many biased estimators such as stain estimator, order ridge estimator, principal components estimator root root estimator and Liu estimator etc. have been developed [1-5], which have successfully applied to many fields. However, these biased estimators are not robust, this is, they are sensitive to observations contaminated by outlier. So robust estimation has been investigated [6-9] such as M estimation, which can overcome the impact of observation outliers. When multicollinearity and outliers coexisting parameter estimation is more difficult, some biased robust estimation are proposed [10-18], which is robust against outliers an well, multicollinearity. However, parameter estimation when high leverage point [19, 20], multicollinearity and outliers are simultaneous, has been nearly ignored. Few works have been reported [21, 22].

The purpose of this paper is to propose a class of new estimators which are robust against multicollinearity and , as well, outliers and high leverage point. After a brief review over the generalized M (GM) estimation in Section 2, we define a class of estimators, generalized shrunken type GM estimators, by rafting the GM estimation technique into biased estimation in Section 3. A number numerical example is presented in section4 to illustrate that these new estimators are better than the biased estimators and the robust estimators when multicollinearity, outliers and high leverage point exist simultaneously.

## 2. Generalized M estimation

Consider the linear regression model of the form

$$\begin{cases} y = X\beta + e \\ E(e) = 0, Cov(e) = \sigma^2 I \end{cases} \quad (1)$$

where $\mathbf{y}=(y_1, y_2,\cdots, y_n)^T$ is an $n\times 1$ vector of response observations, $\mathbf{X}=(\boldsymbol{x}_1,\boldsymbol{x}_2,\cdots,\boldsymbol{x}_n)^T$ is an $n\times p$ matrix of the levels of the regression variables, $\boldsymbol{x}_i$ is the i-th row of X, $\boldsymbol{\beta}$ is $p\times 1$ vector of unknown parameters, $\mathbf{e}=(e_1,e_2,\cdots,e_n)^T$ is an $n\times 1$ vector of random errors, and **I** is $n\times n$ identity matrix. The errors are assumed to have $E(\boldsymbol{e}_i)=0$ and $D(\boldsymbol{e}_i)=\sigma^2$, and assumed to be uncorrelated. The classic least squares (LS) estimates for the model coefficient is given by

$$\hat{\boldsymbol{\beta}}_{LS}=(\boldsymbol{X}^T\boldsymbol{X})^{-1}\boldsymbol{X}^T\boldsymbol{y}$$

Note that GM estimators are the solution of the equations

$$\sum_{i=1}^{n}\psi(\boldsymbol{y}_i-\boldsymbol{x}^T_{(i)}\boldsymbol{\beta})\boldsymbol{x}_{(i)}=0 \tag{3}$$

where $\psi(\bullet)$ is a bounded real-valued function defined on $R^p\times R$. When the weighted functions are $w_i=\psi(r_i/s)/(r_i/s)$, where s is estimated value of $\sigma$, therefore equations (3) is described by

$$\boldsymbol{X}^T\boldsymbol{W}\boldsymbol{r}=\boldsymbol{0}, \tag{4}$$

where $\boldsymbol{r}=\boldsymbol{y}-\boldsymbol{X}\hat{\boldsymbol{\beta}}$ is residual vector and $\boldsymbol{W}=diag(w_1,w_2,\cdots,w_n)$ is weight matrix. So equations (4) can be written in weighted LS form as

$$\hat{\boldsymbol{\beta}}_{GM}=(\boldsymbol{X}^T\boldsymbol{W}\boldsymbol{X})^{-1}\boldsymbol{X}^T\boldsymbol{W}\boldsymbol{y} \tag{5}$$

or

$$\boldsymbol{X}^T\boldsymbol{W}\boldsymbol{y}-\boldsymbol{X}^T\boldsymbol{W}\boldsymbol{X}\boldsymbol{\beta}=\boldsymbol{0} \tag{6}$$

Statistical functional of GM estimation is given by the solution of the equations

$$\int\psi(\boldsymbol{y}-\boldsymbol{X}^T\boldsymbol{\beta})\boldsymbol{X}dF(\boldsymbol{X},\boldsymbol{y})=0 \tag{7}$$

where F is joint distribution function of (**X**, **y**). And influence function of GM estimation is

$$IF((\boldsymbol{X},\boldsymbol{y}),\boldsymbol{\beta}_{GM},F)=\psi(y-\boldsymbol{X}^T\boldsymbol{\beta}_{GM})\boldsymbol{B}_{GM}^{-1}\boldsymbol{X} \tag{8}$$

where $\boldsymbol{B}_{GM}=\int\psi'(\boldsymbol{X},\boldsymbol{y}-\boldsymbol{X}^T\boldsymbol{\beta}_{GM})\boldsymbol{X}\boldsymbol{X}^TdF(\boldsymbol{X},\boldsymbol{y})\cdot\boldsymbol{B}_{GM}$ is generally a positive definite matrix. By letting

$$\psi(\boldsymbol{X},\boldsymbol{y}_i-\boldsymbol{x}^T_{(i)}\boldsymbol{\beta})=\gamma(\boldsymbol{x}_{(i)})\eta(\boldsymbol{y}_i-\boldsymbol{x}^T_{(i)}\boldsymbol{\beta}/\gamma(\boldsymbol{x}_{(i)})) \tag{9}$$

where $\gamma(\boldsymbol{x}_{(i)})=(1-h_{ii})/\sqrt{h_{ii}}$, $h_{ii}$ is the diagonal element of the matrix $\boldsymbol{H}=X(X^TX)^{-1}X^T$ and $\eta(\bullet)$ is a bounded function, then the influence function of GM estimation is function of bounded, when high leverage points are present.

Formally, the expression of the robust estimator (5) is identical to the M estimator, the only difference is that the weight matrix in the M estimation is replaced by the equivalent weight matrix in the GM estimator. It is this substitution that enables the robust estimation capable of resisting the influence of high leverage point, because we can choose a suitable weight matrix that has exactly this ability.

But Gui (2001) derived the asymptotic estimator of variance-covariance matrix for the robust estimator $\hat{\boldsymbol{\beta}}_{GM}$, which is given by

$$\sum_{\hat{\boldsymbol{\beta}}_{GM}}\approx\hat{\sigma}^2(\boldsymbol{X}^T\boldsymbol{W}\boldsymbol{X})^{-1} \tag{10}$$

where

$$\hat{\sigma}^2 = \frac{1}{\text{n - p}} \frac{\sum_{i=1}^{n} w_i \psi^2(r_i)}{\left[\frac{1}{n}\sum_{i=1}^{n} \psi'(r_i)\right]^2} \,. \tag{11}$$

It is easily seen from (10) that the asymptotic variance-covariance matrix of robust estimator is of the same form as the variance –covariance matrix of the LS estimator. In particular, when multicollinearity among the columns of the matrix **X** is present, the matrix $X^T WX$ maybe ill-conditioned, implying relatively large asymptotic variances. Hence, we expect the GM estimator to perform asymptotically, similar to the LS estimator in the presence of multicollinearity. So it becomes necessary to modify the robust estimator in the line of biased estimator.

## 3. Generalized shrunken type GM estimation

In response to the above perceived deficiency with the robust estimator, we propose a new estimator by grafting the biased estimation technique (Massy 1965; Hoerl and Kennard 1970a,1970b; Liu 1993; Gruber 1998) philosophy into the robust estimator, which may be designated the generalized shrunken type GM (GSGM) estimator and be denoted by $\hat{\boldsymbol{\beta}}_{GM}$

$$\hat{\boldsymbol{\beta}}_{GSGM} = \hat{\boldsymbol{\beta}}_{GSGM}(S) = \tilde{\boldsymbol{Q}}\boldsymbol{S}\tilde{\boldsymbol{Q}}^T \hat{\boldsymbol{\beta}}_{GM} = \tilde{\boldsymbol{Q}}\boldsymbol{S}\tilde{\boldsymbol{\Lambda}}^{-1}\tilde{\boldsymbol{Q}}^T \boldsymbol{X}^T\tilde{\boldsymbol{W}}\boldsymbol{y} \tag{12}$$

where $\tilde{Q}$ is an orthogonal matrix satisfying

$$\tilde{\boldsymbol{Q}}^T \boldsymbol{X}^T\tilde{\boldsymbol{W}}\boldsymbol{X}\tilde{\boldsymbol{Q}} = \tilde{\boldsymbol{\Lambda}} = diag(\tilde{\lambda}_1,\cdots,\tilde{\lambda}_p)\,,\ \tilde{\lambda}_1 \geq \cdots \geq \tilde{\lambda}_p > 0.$$

That is the spectral decomposition of matrix $\boldsymbol{X}^T\boldsymbol{W}\boldsymbol{X}$ , and

$$\boldsymbol{S} = diag(s_1,\cdots,s_p)\ \ 0 \leq s_i \leq 1, i = 1,2,\cdots,p)$$

is called a shrinking parameter matrix.

The most common solution approach is iteratively reweighted least squares (IRLS), we assume that the estimator $\hat{\boldsymbol{\beta}}^{(m)}_{GSGM}$ , the residual $\boldsymbol{r}^{(m)} = \boldsymbol{y} - \boldsymbol{X}\hat{\boldsymbol{\beta}}^{(m)}_{GSGM}$ and spectral decomposition of matrix $\tilde{\boldsymbol{Q}}^{(m)T}\boldsymbol{X}^T\tilde{\boldsymbol{W}}^{(m)}\boldsymbol{X}\tilde{\boldsymbol{Q}}^{(m)} = \tilde{\boldsymbol{\Lambda}}^{(m)}$ to the m-th step estimation are known, then we can give the (m+1)-th step estimator

$$\hat{\boldsymbol{\beta}}^{(m+1)}_{GSGM} = \tilde{\boldsymbol{Q}}^{(m)}\boldsymbol{S}^{(m)}\tilde{\boldsymbol{\Lambda}}^{(m)-1}\tilde{\boldsymbol{Q}}^{(m)T}\boldsymbol{X}^T\tilde{\boldsymbol{W}}^{(m+1)}\boldsymbol{y} \tag{13}$$

where

$$\tilde{w}_i^{(m+1)} = \frac{\psi\left(\dfrac{r_i^{(m)}}{\hat{\sigma}^{(m)}_{MAD}\gamma^{(m)}(x_{(i)})}\right)}{\dfrac{r_i^{(m)}}{\hat{\sigma}^{(m)}_{MAD}\gamma^{(m)}(x_{(i)})}} \tag{14}$$

$$r_i^{(m)} = y_i - x_{(i)}^T\hat{\boldsymbol{\beta}}^{(m)}_{GSGM} \tag{15}$$

$$\tilde{\gamma}^{(m)}(x_{(i)}) = \frac{1-\tilde{h}_{ii}^{(m)}}{\sqrt{\tilde{h}_{ii}^{(m)}}} \tag{16}$$

$$\tilde{\boldsymbol{H}}^{(m)} = (\tilde{h}_{ij}^{(m)})_{n\times n} = \boldsymbol{X}(\boldsymbol{X}^T\tilde{\boldsymbol{W}}^{(m)}\boldsymbol{X})^{-1}\boldsymbol{X}^T\tilde{\boldsymbol{W}}^{(m)} \tag{17}$$

$$\hat{\sigma}_{\mathrm{MAD}}^{(m)} = \frac{\operatorname{med}_i\{|r_i^{(m)} - \operatorname{med}_j\{r_j^{(m)}\}|\}}{0.6745} \tag{18}$$

Obviously, the GM estimator $\hat{\boldsymbol{\beta}}_{GM}$ refer to the case where **S**=**I**. Further, according to concrete problems in practical application, we can generate many useful estimators by appropriate of the shrinking parameter matrix **S**. Several important estimators are given next.

*3.1 Ridge type GM estimator*

When $\mathbf{S} = \tilde{\mathbf{\Lambda}}(\tilde{\mathbf{\Lambda}} + k\mathbf{I}_p)^{-1}$, we can get

$$\begin{aligned}\hat{\boldsymbol{\beta}}_{ORGM}(k) &= (\boldsymbol{X}^T\tilde{\boldsymbol{W}}\boldsymbol{X} + k\boldsymbol{I})^{-1}\boldsymbol{X}^T\tilde{\boldsymbol{W}}\boldsymbol{y} \\ &= (\boldsymbol{X}^T\tilde{\boldsymbol{W}}\boldsymbol{X} + k\boldsymbol{I})^{-1}\boldsymbol{X}^T\tilde{\boldsymbol{W}}\boldsymbol{X}\hat{\boldsymbol{\beta}}_{GM}\end{aligned} \tag{19}$$

which is called a ridge type GM (ORGM) estimator, where $k \geq 0$ is called a ridge parameter.

*3.2 Principal components type GM estimator*

When

$$\boldsymbol{S} = \begin{pmatrix} \boldsymbol{I}_r & 0 \\ 0 & 0 \end{pmatrix}$$

we get

$$\hat{\boldsymbol{\beta}}_{PCGM}(r) = \tilde{\boldsymbol{Q}}_r\tilde{\boldsymbol{\Lambda}}_r^{-1}\tilde{\boldsymbol{Q}}_r^T\boldsymbol{X}^T\tilde{\boldsymbol{W}}\boldsymbol{y} = \tilde{\boldsymbol{Q}}_r\tilde{\boldsymbol{Q}}_r^T\hat{\boldsymbol{\beta}}_{GM} \tag{20}$$

which is called a principal-component-type GM (PCGM) estimator, where

$$\tilde{\boldsymbol{Q}}_r = \tilde{\boldsymbol{Q}}\begin{pmatrix} \boldsymbol{I}_r \\ 0 \end{pmatrix}, 1 \leq r \leq p$$

*3.3 Liu type GM estimator*

When $\boldsymbol{S} = (\tilde{\boldsymbol{\Lambda}} + \boldsymbol{I})^{-1}(\tilde{\boldsymbol{\Lambda}} + d\boldsymbol{I})\boldsymbol{\Lambda}^{-1}$, we get

$$\hat{\boldsymbol{\beta}}_{\mathrm{LGM}}(d) = \tilde{\boldsymbol{Q}}(\tilde{\boldsymbol{\Lambda}} + \boldsymbol{I})^{-1}(\tilde{\boldsymbol{\Lambda}} + d\boldsymbol{I})\boldsymbol{\Lambda}^{-1}\tilde{\boldsymbol{Q}}^T\boldsymbol{X}^T\tilde{\boldsymbol{W}}\boldsymbol{y} = (\boldsymbol{X}^T\tilde{\boldsymbol{W}}\boldsymbol{X} + \boldsymbol{I})^{-1}(\boldsymbol{X}^T\tilde{\boldsymbol{W}}\boldsymbol{X} + d\boldsymbol{I})\tilde{\boldsymbol{X}}_{GM} \tag{21}$$

which is called a Liu type GM (LGM) estimator, where $0 < d < 1$.

It is also an important problem in application that selection ways of partial parameters above estimators, according to ways of selected partial parameters in biased estimators, we propose two ways to each estimator.

Ridge parameter k of ORGM estimator is obtained by

$$\hat{k} = \frac{p\hat{\sigma}_{MAD}^2}{\hat{\boldsymbol{\beta}}_{ORGM}^T\boldsymbol{X}^T\boldsymbol{X}\hat{\boldsymbol{\beta}}_{ORGM}} \quad \text{or} \quad \hat{k} = \frac{p\hat{\sigma}_{MAD}^2}{\hat{\boldsymbol{\beta}}_{ORGM}^T\hat{\boldsymbol{\beta}}_{ORGM}}$$

Principal component number r of PCGM estimator is obtained by

$$\tilde{\lambda}_r >> \tilde{\lambda}_{r+1} > 0.001 \quad \text{or} \quad \sum_{i=1}^{r}\tilde{\lambda}_i / \sum_{i=1}^{p}\tilde{\lambda}_i \geq 75\% \text{ or } 80\%$$

Partial parameter d of LGM estimator is obtained by

$$\hat{d}=\frac{\sum_{i=1}^{p}\frac{\hat{\alpha}_i^2-\hat{\sigma}_{MAD}^2}{(\tilde{\lambda}_i+1)^2}}{\sum_{i=1}^{p}\frac{\tilde{\lambda}_i\hat{\alpha}_i^2+\hat{\sigma}_{MAD}^2}{\tilde{\lambda}_i(\tilde{\lambda}_i+1)^2}} \quad \text{or} \quad \hat{d}=1-\frac{\hat{\sigma}_{MAD}^2\sum_{i=1}^{p}\frac{1}{\tilde{\lambda}_i(\tilde{\lambda}_i+1)^2}}{\sum_{i=1}^{p}\frac{\hat{\alpha}_i^2}{(\tilde{\lambda}_i+1)^2}}$$

where $\hat{\alpha}_{GM}=(\hat{\alpha}_1,\cdots,\hat{\alpha}_p)^T=\tilde{Q}^T\hat{\beta}_{GM}$ and $\tilde{Q}^TX^T\tilde{W}X\tilde{Q}=\tilde{\Lambda}=diag(\tilde{\lambda}_1,\cdots,\tilde{\lambda}_p)$.

## 4. Numerical example

In order to illustrate the performance of the proposed GSGM estimators, we consider a set of data that reflect the information take from Myers (1990) as shown in table 1. They are 17 hospitals human resource data of U.S. Navy at various sites around the world. There are five regressors and one response variable. The aim here is to produce an empirical equation that will estimate manpower needs for Naval hospitals. Where y is average monthly workload; x1 is the average number of patients is received a daily; x2 is average number of the x-ray is used; x3 is bed occupancy days per month; x4 is the eligible population in local district /1000; x5 is the average length of the patients stay in hospital.

The correlation matrix and condition number (77593) indicate that there is a serious linear dependency among the regressors, Therefore we should use ridge estimators to reduce the effect of the multicollinearity on the parameter estimates.

Further analysis of the data indicates that the cases 6,10,14,16 and 17 are identified as outlier, and the cases 10,15,16 and 17 have very high leverage. These analyses give that the dataset has simultaneously multicollinearity and outliers in both direction.

**Table 1.** Hospital manpower resource data.

| Site | X1 | X2 | X3 | X4 | X5 | y |
|---|---|---|---|---|---|---|
| 1 | 15.57 | 2463 | 472.92 | 18.0 | 4.45 | 566.52 |
| 2 | 44.02 | 2048 | 1339.75 | 9.5 | 6.92 | 696.82 |
| 3 | 0.42 | 3940 | 620.25 | 12.8 | 4.28 | 1033.15 |
| 4 | 18.74 | 6305 | 568.33 | 36.7 | 3.90 | 1603.62 |
| 5 | 49.20 | 5723 | 1497.6 | 35.7 | 5.50 | 1611.37 |
| 6 | 5.48 | 11520 | 1365.83 | 24.0 | 4.60 | 1613.27 |
| 7 | 59.28 | 5579 | 1687.00 | 43.3 | 5.62 | 1854.17 |
| 8 | 94.39 | 5969 | 1639.92 | 46.7 | 5.15 | 2160.55 |
| 9 | 128.02 | 8461 | 2872.33 | 78.7 | 6.18 | 2305.58 |
| 10 | 128.02 | 20106 | 3655.08 | 180.5 | 6.15 | 3503.93 |
| 11 | 96.00 | 133131 | 2912.00 | 60.9 | 5.88 | 3571.89 |
| 12 | 13.42 | 10771 | 3921.00 | 103.7 | 4.88 | 3741.40 |
| 13 | 127.21 | 15543 | 3865.67 | 157.7 | 7.00 | 10343.81 |
| 14 | 252.90 | 36194 | 7684.10 | 157.7 | 7.00 | 10343.81 |
| 15 | 409.20 | 34703 | 12446.33 | 169.4 | 10.78 | 11932.17 |
| 16 | 463.70 | 39204 | 14098.40 | 331.4 | 7.05 | 15414.94 |
| 17 | 510.22 | 86533 | 15524.00 | 371.6 | 6.35 | 18854.45 |

Since the real value of parameters are unknown, we will measure the accuracy of estimations by forecast period mean squared error (MSEP) (Chen 1987). First we select any one as forecast objectives except for the cases 6,10,14,15,16 and 17, then to compute estimators of parameter, at last, to calculate the MSEP. According to this criterion, we consider the LS estimators , ridge estimators, M estimators, ridge type M (ORM) estimators, Principal components type M (PCM) estimators, Liu type M (LM) estimators, ORGM estimators, PCGM estimators and LGM estimators of parameter, then calculate their MSEP separately in Table 2.

**Table 2.** The MSEP of various estimators.

| estimation | LS | R | M | ORM | PCM | LM | ORGM | PCGM | LGM |
|---|---|---|---|---|---|---|---|---|---|
| MSEP | 356.2 | 277.8 | 271.3 | 173.7 | 265.0 | 80.2 | 99.6 | 74.4 | 131.3 |

It is obvious to see that MSEP of GSGM estimation is most little in all kinds of estimators in Table 2 it illuminates that GSGM estimators can not only overcome the multicollinearity of coefficient matrix and outliers but also have the ability to control the influence of leverage points.

## 5. Acknowledgments

Natural Science Foundation of China, NO.41174005，40974009; Foundation and Advanced Technology Research Program of Henan Province, NO.082300410240.

## References

[1] Gruber M H J 1998 *Improving Efficiency by Shrinkage: The James Stein and Ridge Regression Estimators* (NewYork: Marcel Dekker)
[2] Hoerl A E and Kennard R W 1970a Ridge Regression: Biased Estimation for Non-orthogonal Problems *Technometrics* **12** 55
[3] Hoerl A E and Kennard R W 1970b Ridge Regression: Applications to Non-orthogonal Problems *Technometrics* **12** 69
[4] Massy W F 1965 Principal components regression in exploratory statistical research *JASA* **60** 234
[5] Liu K A 1993 New class of biased estimate in linear regression *CSTM* **22**(2) 393
[6] Huber P J 1977 *Robust Covariances Statistical Decision Theory and Related Topics II* ed Gupta and Moore (New York: Academic Press) pp 165
[7] Huber P J 1981 *Robust Statistics* (New York: Wiley)
[8] Hampel F R 1973 Robust Estimation: A Condensed Partial Survey *ZWVG* **27** 87
[9] Maronna R A 1976 Robust M-estimators of Multivariate Location and Scatter *Annals of Statistics* **4** 51
[10] Holland P W 1973 Weighted Ridge Regression: Combining Ridge and Robust Regression Methods *NBER Working Paper Series* Working Paper #11
[11] Pariente S and Welsch R E 1977 Ridge and Robust Regression Using Parametric Linear Programming *Working Paper MIT Alfred P Sloan School of Management*
[12] Askin R G and Montgomery D C 1980 Augmented Robust Estimators *Technometrics* **22** 333
[13] Pfaffenberger R C and Dielman T E 1985 A Comparison of Robust Ridge Estimators *BESPASA* 631
[14] Slivapulle M J 1991 Robust Ridge Regression Based on an M-Estimator *ARJS* **33**(3) 319
[15] Gui Q M, Duan Q T, Zhou Q Y and Guo J F 2001 A New Class of Biased Linear Estimators in Deficient-rank Linear Models *CJM* **16**(1) 71
[16] Xia J L 1988 A Kind of New Biased Estimator of Regression Parameters: Root Estimator *CJMSAP* **3** 21
[17] Nyquist H 1985 *Ridge Type M-estimators LSITCWK* (Berlin: Springer) p 246
[18] Olcay A and Nedret B 2000 Robust Liu estimator for regression based on an M-estimator *JAS* **27** 1

[19] Hampel F R, Ronchetti E M, Rousseeuw P J and Stahel W A 1986 *Robust Statistics: The Approach Based on Influence Function* (New York: John Wiley)
[20] Rousseeuw P J and Zomeren B C 1990 Unmasking Multivariate Outliers and Leverage Points *JASA* **85** 633
[21] Olcay A and Nedret B 1996 Roust Ridge Regression Estimation Biased on the GM-estimators *JM&CS* **22**(3) 1
[22] Simpson J R and Montgomery D C 1996 A biased-robust regression technique for the combined outlier-multicollinearity problem *JSCS* **56**(1) 1
[23] Myers R H 1990 *Classical and Modern Regression with Applications* second edition (Boston: Duxbury Press)
[24] Chen X R and Wang S G 1987 *Advanced Regression Analysis* (Hefei: Anhui Education Publishing House)

IOP Conf. Series: Materials Science and Engineering **57** (2014) 012011 doi:10.1088/1757-899X/57/1/012011

# Artificial neural networks in predicting current in electric arc furnaces

**M Panoiu[1], C Panoiu[1], A Iordan[1] and L Ghiormez[2]**

[1] Polytechnic University of Timisoara – Electrical Engineering and Industrial Informatics Department – Revolutiei street, no. 5, 551128, Hunedoara, Romania
[2] Polytechnic University of Timisoara – Faculty of Automation and Computers – Vasile Parvan street no. 2, 300223, Timisoara Romania

E-mail: manuela.panoiu@upt.ro

**Abstract**. The paper presents a study of the possibility of using artificial neural networks for the prediction of the current and the voltage of Electric Arc Furnaces. Multi-layer perceptron and radial based functions Artificial Neural Networks implemented in Matlab were used. The study is based on measured data items from an Electric Arc Furnace in an industrial plant in Romania.

## 1. Introduction
The Electric Arc Furnace (EAF) is used in the melting process of the scraps which are loaded in the furnace tank. The EAF is one of the modern ways of making steel, but having very large power input ratings. It also behaves like a nonlinear load and can cause large power quality problems, mainly harmonics, interharmonics, the flicker phenomenon and voltage unbalances. Because of these, solutions must be identified through which these disturbances should not be injected in the power supply network. So, the prediction of the future values of the current arc and voltage is need. Based on these values actions can be taken to mitigate (lessen) the disturbances.

## 2. The AC Electric Arc
The power supply equivalent diagram for a single phase is show in figure 1, where r and L being the resistance, respectively the equivalent inductance of the supplying circuit, and $R_A$ and $v_A$ the resistance of the electric arc, respectively the arc voltage [1]. The waveforms of the electrical issues (current arc, voltage arc and voltage supply) are also show in figure 1.

## 3. Modelling the AC Electric Arc
EAFs are used in the production of steel and other metals. EAFs are very large, dynamic and time varying loads due to the nonlinearity of the electric arc. Electric-arc steelmaking generally has two successive stages, a melting stage and a refining stage. For these reasons it is very difficult to find an accurate model for the electric arc. In the reference literature many modeling methods can be found: nonlinear resistive models [3], [4] voltage source models [5], [6] models based on V-I characteristics of the EAF [7], models based on relations between the length of the arc, the voltage and the current of the arc [2], model that used chaos theory [8] and advanced models that use neural networks and fuzzy logic [8-12]

In this paper, in order to model the electric arc, two types of Artificial Neural Networks (ANN) were used: multi-layer perceptron ANN and radial based functions ANN. In order to develop a model based on ANN, actual recorded data from the electrical installation of an EAF are necessary. These data are obtained by measuring the voltages and currents in the secondary side of furnace transformer of an EAF with the capacity of 100 tones [1]. The frequency of acquisition was 5 kHz. The measured data, both from the melting stage and from the refining stage, has been used for ANN training.

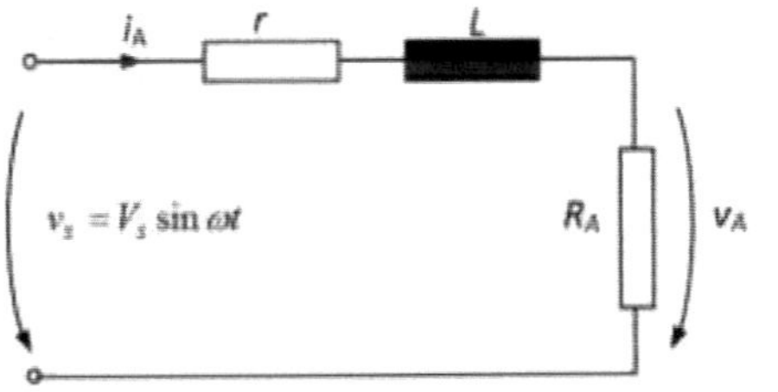


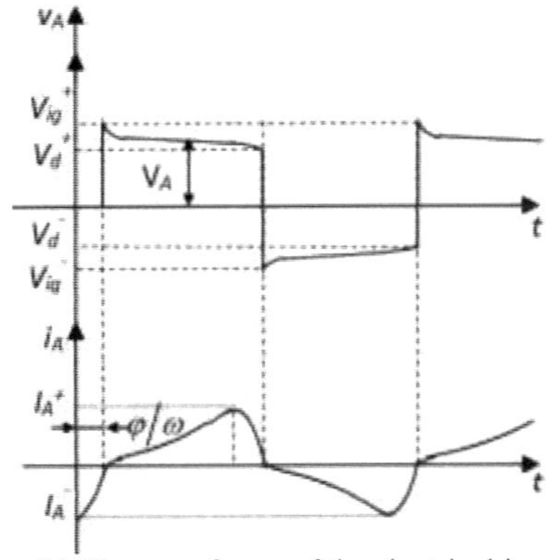


**Figure 1.** a) The power supply equivalent diagram b) The waveforms of the electrical issues

*3.1. Model based multi-layer-perceptron ANN*

A multi-layer-perceptron ANN is a feedforward artificial neural network with one input layer, one output layer and one hidden layer. In figure 2 a multi-layer-perceptron ANN is shown.

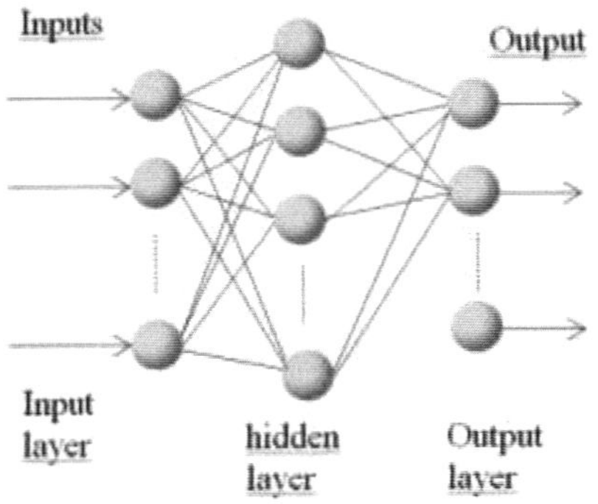


**Figure 2.** Architecture of a multi-layer-perceptron ANN

ANN architecture affects the performance of electric arc modelling. In [13] a study regarding the influence of the ANN architecture to the electric arc model in Matlab was performed. The best results were obtained for a number of neurons in the hidden layer greater than 100. But by over-increasing the number of neurons, the validation and the testing performance of the neural network begin to decrease because the network loses its ability to generalize (overtraining phenomenon).

Based on this study, the architecture of the multi-layer-perceptron was chosen with 80 hidden neurons and a learning rate of 0.8. Other training parameters are: the training function was trainlm, from the Matlab Neural Networks toolbox, trainParam.max_fail = 6, trainParam.epochs = 500. For these values the simulations results for the melting stage are presented in fig. 3 and the ones for the refining stage in figure 4. Figure 3a presents the variation for the measured and simulated voltages, figure 3b presents the variation for the measured and simulated currents and figure 3c presents the voltage – current characteristics of the electric arc for both measured and simulated values (for melting stage). In figures 4a, 4b and 4c similar results are shown but for the refining stage. The results are shown for a single phase in order to obtain a clearer image. By analysing the results, it can be concluded that, in the case of the currents, both for the melting and the refining stage, the prediction is better.

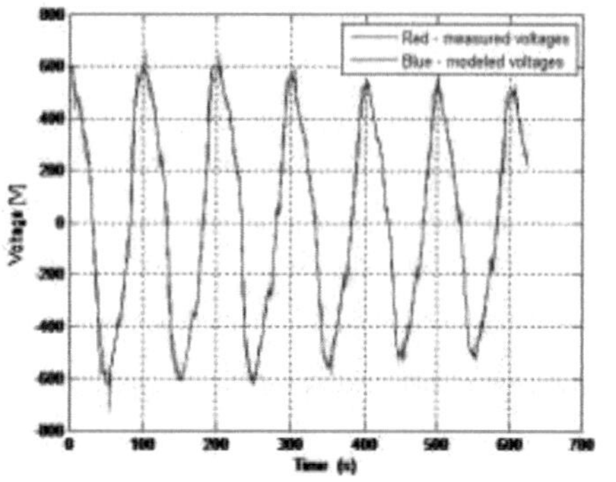


a)

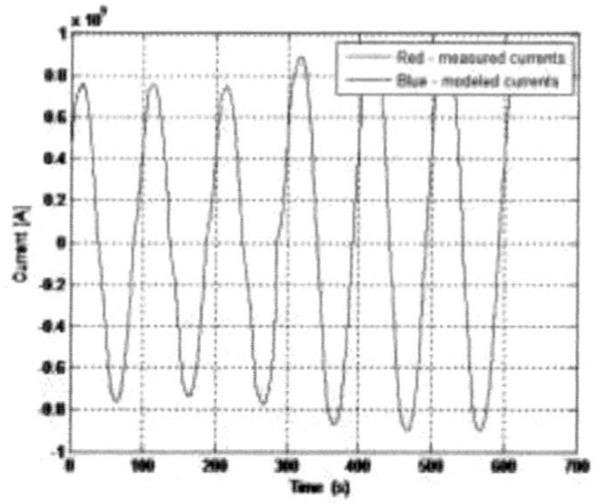


b)

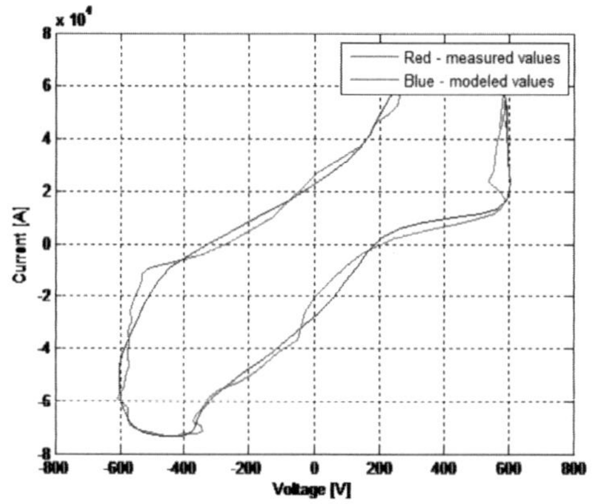


c)

**Figure 3.** Comparisons between measured and simulated items for the melting stage
a) Variation of measured and simulated voltages b) Variation of measured and simulated currents
c) The current – voltage characteristics (measured and simulated)

**Table 1.** The Total Harmonics distortion and the pondered Total Harmonic Distortion for measured and simulated data set (from figure 3).

| **Electric parameter** | | **Measured values** | **Simulated values** |
|---|---|---|---|
| Thdi | Phase 1 | 3.03% | 9,70% |
| | Phase 2 | 2.04% | 6,46% |
| | Phase 3 | 2.18% | 14,50% |
| Thdu | Phase 1 | 8.36% | 6,27% |
| | Phase 2 | 7.92% | 10,54% |
| | Phase 3 | 9.04% | 8,72% |
| Thdpi | Phase 1 | 7.08% | 20,05% |
| | Phase 2 | 4.59% | 13,98% |
| | Phase 3 | 5.22% | 31,24% |
| Thdpu | Phase 1 | 20.59% | 13,74% |
| | Phase 2 | 20.02% | 22,22% |
| | Phase 3 | 26.55% | 20,20% |

For the analysed data, the main electrical parameters of the power quality are also determined and presented in table 1. Hence, the total harmonic distortions of the current and the voltage for all three phases are determined.

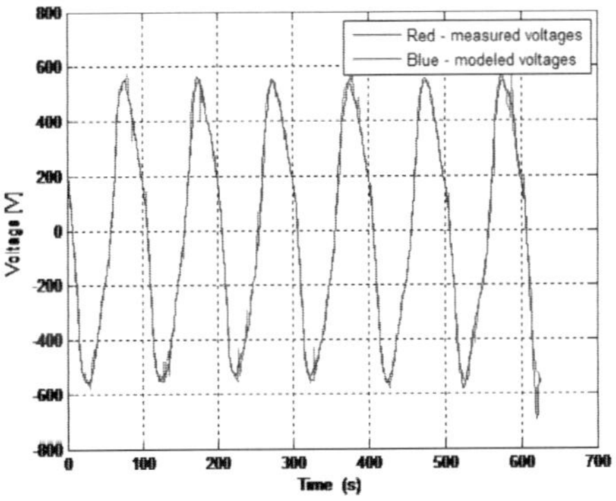


a)

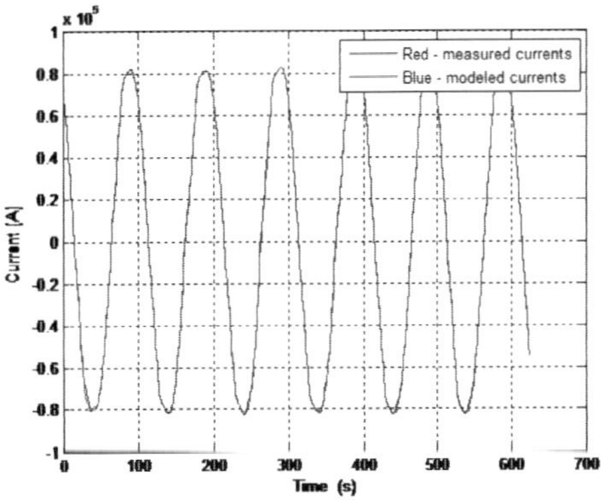


b)

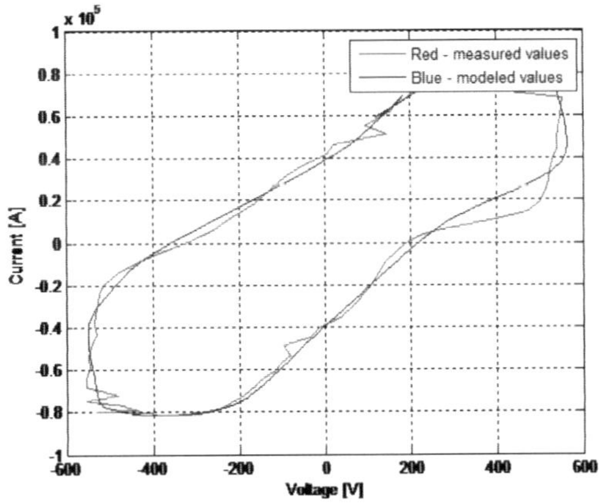


c)

**Figure 4.** Comparisons between measured and simulated items for refining stage
a) Variation of measured and simulated voltages b) Variation of measured and simulated currents
c) The current – voltage characteristics (measured and simulated)

### 3.2. *Model based on radial bases functions ANN*

The radial-basis function artificial neural network (RBF-ANN) is an alternative to the multi-layer perceptron and has gained attention in many applications in recent years due to its simpler structure [12], [14].

RBF-ANN is a feedforward neural network with a much simpler structure than MLP-ANN. A typical structure of RBF-ANN is shown in Figure 5, including the input layer, the hidden layer, and the output layer [14]. The input-output connection consists of two transformations: a nonlinear transformation from the input layer to the hidden layer and linear transformation between hidden and output layers. The RBF networks learn supervised with increasing rates, being designed as functional approximation methods ("curve fitting").

RBF-ANN generally are used for the approximation of curves in multidimensional space. Learning is equivalent to finding a surface in a multidimensional space to match (fit) as described by the input data.

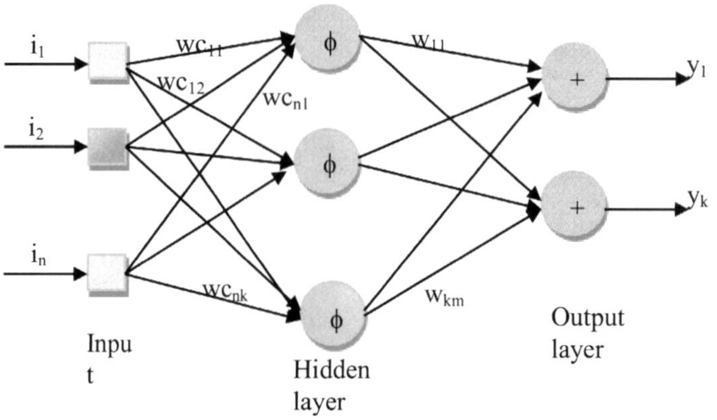


**Figure 5.** RNF-ANN architecture

Compared with the multi-layer perceptron networks (RNA MLP) RBF ANN may require more neurons but their trainning requires less time. This is because the output neurons of the hidden layer MLP are significant for large regions of the input space while the radial functions based on neurons respond only to relatively small regions of the input space. RNA RBF responds better when more vectors are available.

The most common radial basis function used in RBF-ANN is given by

$$\phi_k(x) = \exp\left\{-\frac{(x - wc_k)^T (x - wc_k)}{2\sigma_k^2}\right\}$$

Where x is the input vector data, $w_{ck}$ is central vector for a hidden neuron, $_k$ is the normalization factor.

By using the measured values of the current and of the voltage at the RBF-ANN input as input vectors, respectively estimated values, an approximation for the current voltage characteristic V-I can be obtained.

The training parameters used by newrb function from the Matlab Neural Network Toolbox were eg=0.1 (Mean squared error goal) and spread = 2 (Spread of radial basis functions).

In figures 6 and 7 the results of the simulation are shown.

## 4. Conclusions

In this paper two ANN architectures for the modelling of the electric arc were studied: the multi-layer perceptron ANN (MLP-ANN) and the radial based functions ANN (RBF-ANN). The ANN models were implemented in Matlab 2012 environment. The training and testing results obtained were used to implement an ANN model that can estimate the voltage – current characteristic for both stages of an EAF process: the meltdown stage and the refining stage. The reliability of the predicted currents not only depends on the ANN architecture but also on the input data. Both qualitative and quantitative comparisons were made for the electrical parameters between the simulation results and the performed measurements. By observing the results, it is noticeble that the current waveforms are closely matched by using these advanced models (based on ANN) due to the direct usage of currents as the inputs of the models

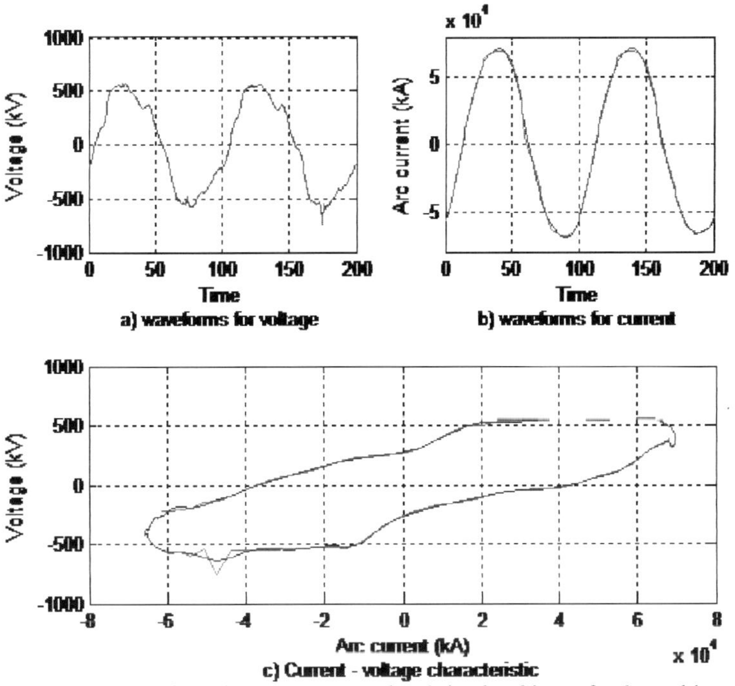


**Figure 6.** Comparisons between measured and simulated items for the melting stage

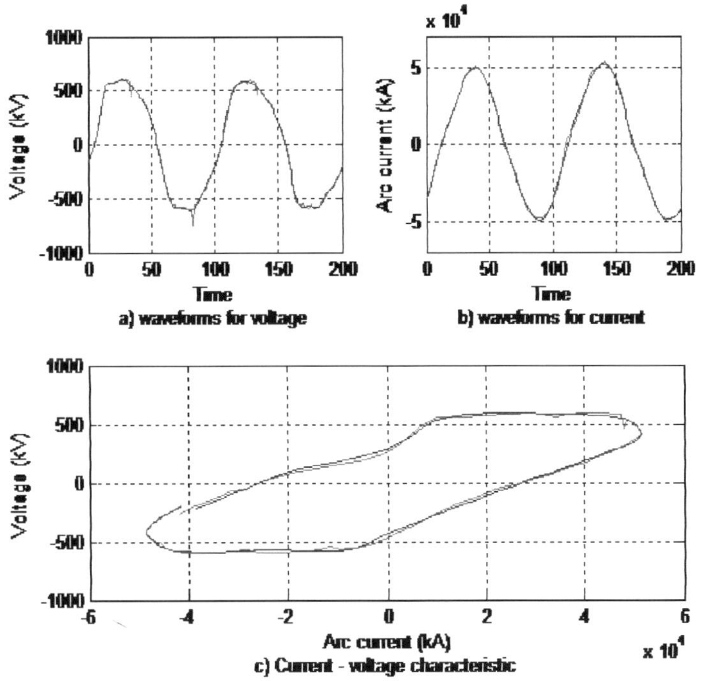


**Figure 7.** Comparisons between measured and simulated items for the refining stage

## References

[1] Panoiu M, Panoiu C and Sora I 2006 Experimental research concerning the electromagnetic pollution generated by the 3-phase electric arc furnaces in the electric power supply networks *Acta Electrotehnica* **47** 102

[2] Panoiu M, Panoiu C and Ghiormez L 2013 Modeling of the Electric Arc Behavior of the Electric Arc Furnace *Advances in Intelligent Systems and Computing* **195** 261

[3] Varadan S, Makram E B and Girgis A A 1996 A new time domain voltage source model for an arc furnace using EMTP *IEEE Trans. Power Deliv.* **11(3)** 1685

[4] Alonso M A P and Donsion M P 2004 An improved time domain arc furnace model for harmonic analysis *IEEE Trans. Power Deliv.* **19(1)** 367

[5] Tang L, Kolluri S and McGranaghan M F 1997 Voltage flicker prediction for two simultaneously operated AC arc furnaces *IEEE Trans. Power Deliv.* **12(2)** 985

[6] Montanari G C, Loggini M, Cavallini A, Pitti L and Zaninelli D 1994 Arc-furnace model for the study of flicker compensation in electrical networks *IEEE Trans. Power Deliv.* **9(4)** 2026

[7] Cano Plata E A and Tacca H E 2005 Arc Furnace Modeling in ATP-EMTP *The 6th International Conference on Power Systems Transients (Montreal, Canada, 20 June 2005)* pp 19-23

[8] Wang F, Jin Z and Zhu Z 2005 Modeling and Prediction of Electric Arc Furnace Based on Neural Network and Chaos Theory *Advances in Neural Networks, Lecture Notes in Computer Science* **3498** 819

[9] King P E and Nyman M D 1996 Modeling and control of an electric arc furnace using a feedforward artificial neural network *Journal of Applied Physics* **80** 1872

[10] Sadeghian A and Lavers J D 2011 Dynamic reconstruction of nonlinear v–i characteristic in electric arc furnaces using adaptive neuro-fuzzy rule-based networks *Applied Soft Computing* **11** 1448

[11] Chang G W Cheng C I and Liu Y J 2010 A neural-network-based method of modeling electric arc furnace load for power engineering study *IEEE Transactions on Power Systems* **25** 138

[12] Chang G W Liu Y J and Chen C I 2008 Modeling voltage-current characteristics of an electric arc furnace based on actual recorded data: A comparison of classic and advanced models *IEEE Power and Energy Society General Meeting - Conversion and Delivery of Electrical Energy in the 21st Century (Pittsburgh, USA, 20-24 July 2008)* pp 1-6

[13] Panoiu M, Panoiu C, Ghiormez L, Iordan A 2013 A numerical Modeling of Nonlinear Behavior Using Artificial Neural Networks *AIP Conference Proceedings (Rhodes, Greece, 22-27 September 2013)* pp 1337-1340

[14] Chang G W, Cheng I C, Yu-Feng T 2010 Radial-Basis-Function-Based Neural Network for Harmonic Detection *IEEE Transactions On Industrial Electronics* **57(6)** 2171

# A new approach in the design of an interactive environment for teaching Hamiltonian digraphs

**A E Iordan[1], M Panoiu[1]**

[1] Electrical Engineering and Industrial Informatics Department, Engineering Faculty of Hunedoara, Polytechnic University of Timisoara

E-mail: anca.iordan@fih.upt.ro

**Abstract.** In this article the authors present the necessary steps in object orientated design of an interactive environment that is dedicated to the process of acquaintances assimilation in Hamiltonian graphs theory domain, especially for the simulation of algorithms which determine the Hamiltonian trails and circuits. The modelling of the interactive environment is achieved through specific UML diagrams representing the steps of analysis, design and implementation. This interactive environment is very useful for both students and professors, because computer programming domain, especially digraphs theory domain is comprehended and assimilated with difficulty by students.

## 1. Introduction

Graphs theory [1] represents the mathematic study of abstract relationship structure between the objects. Although graphs are themselves purely theoretical, their ability to model pair-wise relationships in systems of arbitrary complexity yields abundant direct correspondence with numerous important physical systems in the real world. More than this, the fundamental graph structure permits a geometrical view of them in many ways. If they are taken together, these two properties suggest that graph theory teaching and also researching in this domain can obtain benefits using an interactive environment addressed Hamiltonian graphs study.

Starting with this necessity, we designed an interactive environment that can be used by students and also by graphs theory beginners, but strong enough to help the researchers in graphs theory domain. In educational scope, the interface and the visual components will lead the students and the beginners in this domain to a better comprehending of the next algorithms using for Hamiltonian trails and circuits [2] determination: Chen algorithm [3], Kaufmann algorithm [4], Foulkes algorithm [5] and Albertson algorithm [6].

## 2. Interactive Environment Development

Using UML, interactive environment analysis consists in accomplishment of use case diagram [7]. The interactive environment is described in a comprehensible and concise manner as representing the use cases. Each case specifies the interactions among users and environment. UML uses case diagram is created in an iterative approach (figure 1). Diagram performed defines the environment domain, allowing visualization of the size and sphere of the action for the entire realization process. This includes:

- Three actors - the users that have external entities with which the environment interacts;
- Seven use cases that describe the functionality of the interactive environment;
- Relationships between users and use cases (association relationships), relationships between use cases (dependency and generalization relationships), and relationships between actors.

Oriented object methodologies [8] introduced the representation of the static structure of application using classes and relations between them. Through oriented object modelling of the Java application have been identified classes and relationships of inheritance and realization which are presented in figure 2.

- To memorize the vertices of the graph has been implemented "Vertex" class.
- To memorize the arcs of the graph has been implemented "Arc" class.
- To memorize a graph has been implemented "Graph" class.
- For the drawing of a Hamiltonian trail or circuit has been implemented "Hamiltonian" class.
- For the drawing of a graph has been implemented "GraphRepresentation" class.
- For the determination of the Hamiltonian trail using Chen algorithm has been implemented "HamiltonianChen" class.
- For the determination of the Hamiltonian trails or circuits using Albertson algorithm has been implemented "HamiltonianAlbertson" class.

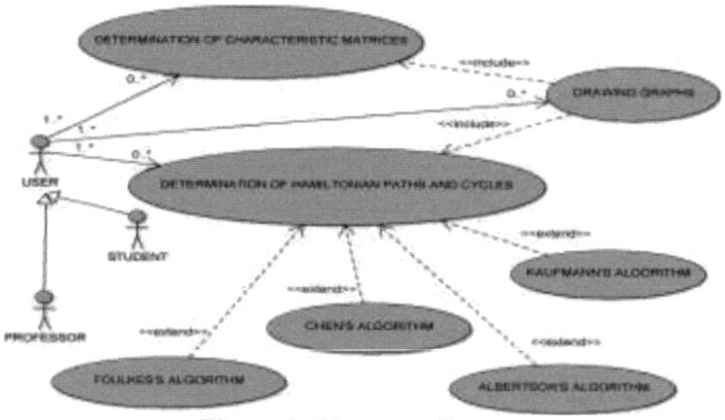


**Figure 1.** Use cases diagram

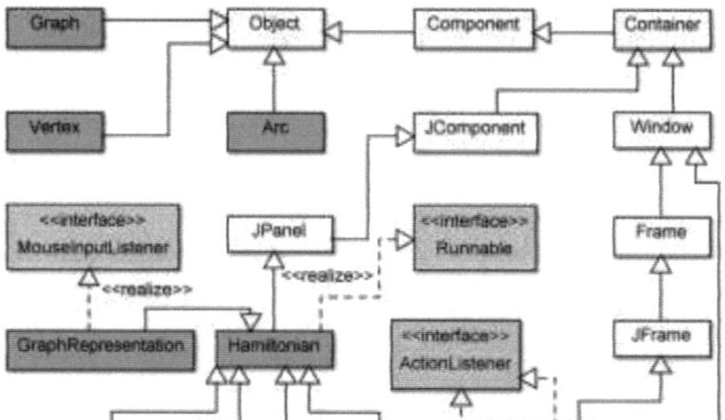


**Figure 2.** Diagram that presents relationships between classes

- For the determination of the Hamiltonian trails or circuits using Kaufmann algorithm has been implemented "HamiltonianKaufmann" class.
- For the determination of the Hamiltonian trails or circuits using Foulkes algorithm has been implemented "HamiltonianFoulkes" class.
- In order to achieve the window that will form graphical user interface of the application has been implemented "Algorithm" class.
- To simulate Chen algorithm has been implemented "Chen" class.
- To simulate Albertson algorithm has been implemented "Albertson" class.
- To simulate Foulkes algorithm has been implemented "Foulkes" class.
- To simulate Kaufmann algorithm has been implemented "Kaufmann" class.
- To introduce the weight of an arc has been implemented "Weight" class.

We can observe that the "Algorithm" class inherit attributes and methods of the "JFrame" class, but implements the "ActionListener" interface. "Weight" class inherits attributes and methods of the "JDialog" class, but implements the interface "ActionListener". "Hamiltonian" class inherit attributes and methods of the "Jpanel" class, but implements "Runnable" interface, and "GraphRepresentation" class inherits attributes and methods of the "Hamiltonian" class and implements the "MouseInputListener" interface.

Between instances of the classes presented in figure 2 there are especially composition and aggregation relationships. In the composition relationship, unlike the aggregation relationship, the instance cannot exist without the party objects. Analysing figure 3 we can observe that an instance of "Hamiltonian" type consists in two objects of "Graph" type, one of "Graphics2D" type and the other of "Thread" type. Aggregation relationship is an association where it's specified who is integer and who is a part. For example, an object of "Arc" type or represent a part from an object of "Graph" type.

Collaboration diagrams describe the behaviour of a set of objects in a certain context with an emphasis on organizing the objects involved in the interaction. Diagram presented in figure 4 renders the interactions between objects that allow determining the Hamiltonian trail using Foulkes algorithm.

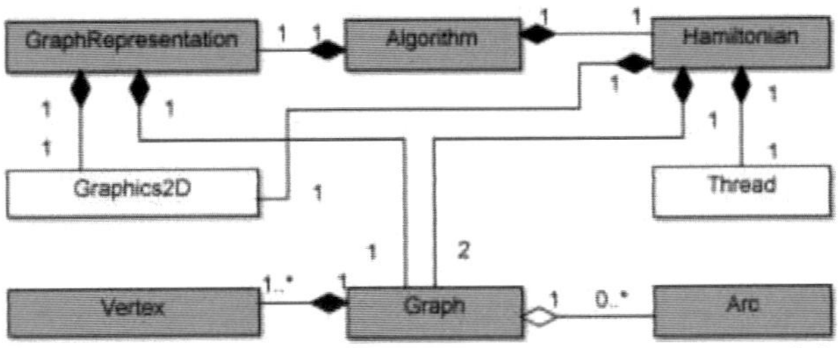


**Figure 3.** Diagram that presents relationships between instances of classes

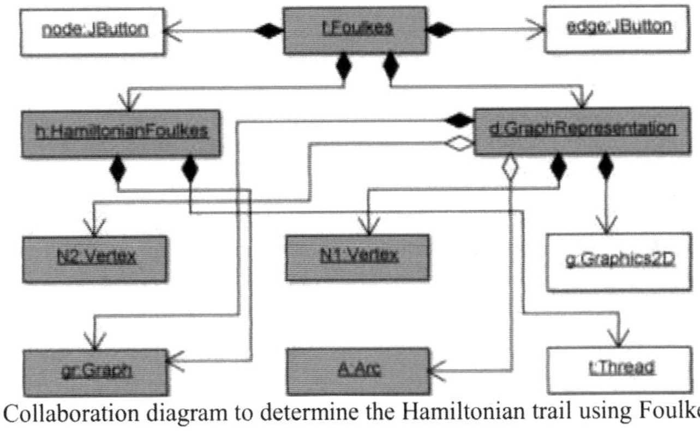


**Figure 4.** Collaboration diagram to determine the Hamiltonian trail using Foulkes algorithm

Component diagram [9] is analogue to packages diagram, allowing visualization of how the interactive environment is divided and the dependencies among class modules. The diagram presented in figure 5 describes the set of components that together provide environment performance.

Central component of the diagram is "Algorithm.class", a component obtained by transforming by the Java compiler into executable code of the "Algorithm.java" component. As can be seen that component interacts directly with components "GraphRepresentation.class", "Hamiltonian.class", "Kaufmann.class", "Foulkes.class", "Chen.class" and "Albertson.class".

Component "Hamiltonian.class" that is obtained by Java compiler from component "Hamiltonian.java" in executable code interactions directly with components "Graph.class", "GraphRepresentation.class", "HamiltonianAlbertson.class", "HamiltonianKaufmann.class", "HamiltonianFoulkes.class" and "HamiltonianChen.class".

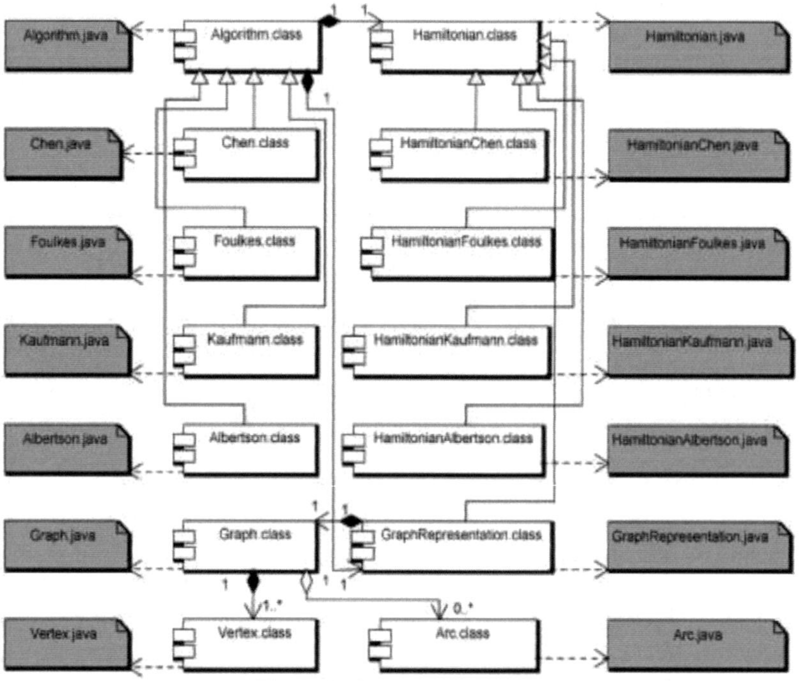


**Figure 5.** Component diagram

## 3. Interactive environment graphical interface

The interactive environment was implemented in Java [10] as independent application. The application can easily convert in a Java applet. By using visual simulations in computer assisted learning the efficiency of learning is increased.

Starting from particularized requirements in use case diagram (figure 1) it was designed graphical user interface of the interactive environment that contains a bar with three menus as in figure 6. In the same figure is represented the Hamiltonian trail obtained into a digraph with twelve vertices and four strongly connected components.

First menu contains four options:

- New graph – permits creating a new directed graph by specifying the vertices and arcs using the mouse.
- Load graph – permits graphical representation of a graph read from an existing file.
- Save graph – permits saving the information about a current graph.

- Exit – permits to exit from an application, any unsaved digraph being lost.

The second menu contains five options that will permit determination of adjacency matrix, incidence matrix, distance matrix, Latin matrix and reduced Latin matrix for the current graph. The final menu, "Hamiltonian Trails", contains four commands and permits determination of the Hamiltonian trails and circuits in current graph only if this one exists, using one of the following four algorithms: Chen algorithm, Kaufmann algorithm, Albertson algorithm or Foulkes algorithm.

For the exemplifications of the simulation, is presented Chen algorithm. In figure 7 are presented steps that are made in determination of the Hamiltonian trail in a digraph with 8 vertices without circuits. Corresponding to Chen theorem [3], if the Hamiltonian trail exists, then it is unique. Figure 8 presents the obtained Hamiltonian trail for the digraph used for exemplification.

Hereinafter is simulated Kaufmann algorithm [4], also called Latin multiplication algorithm. This algorithm uses two matrices: the Latin matrix and the reduced Latin matrix.

For the digraph with 5 vertices which is drawn in window presented in figure 9, in which can be observed these matrices by selecting corresponding options of the second menu. In figure 10 are presented one Hamiltonian trail and five Hamiltonian circuits obtained for the digraph with 5 vertices. The window in which are presented these results is represented in figure 11.

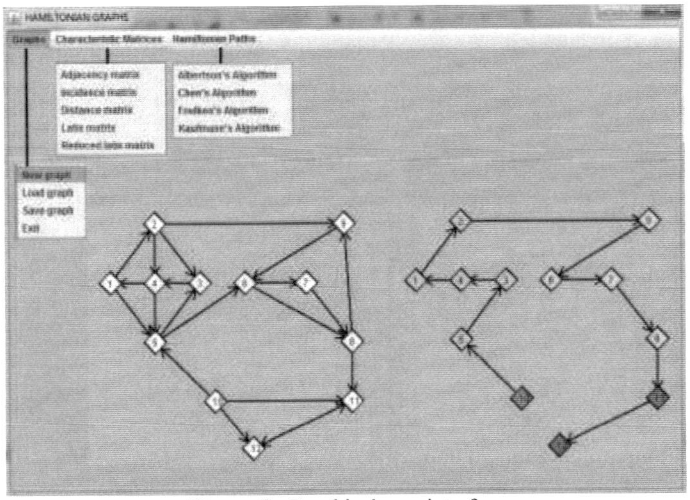


**Figure 6.** Graphical user interface

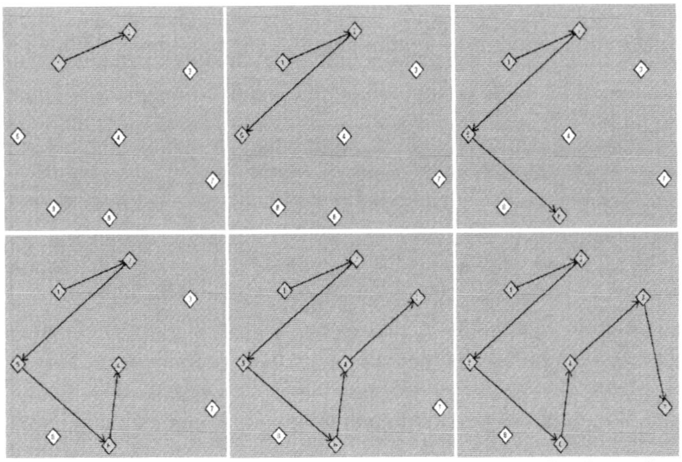

**Figure 7.** Implementation steps in determining the Hamiltonian trail

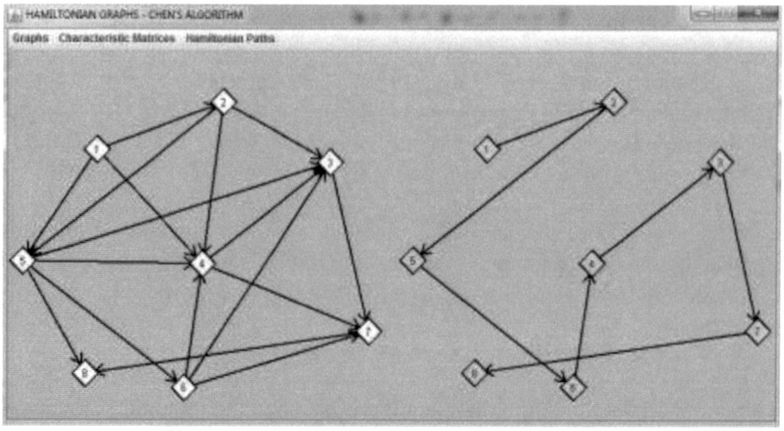

**Figure 8.** Hamiltonian trail using Chen algorithm

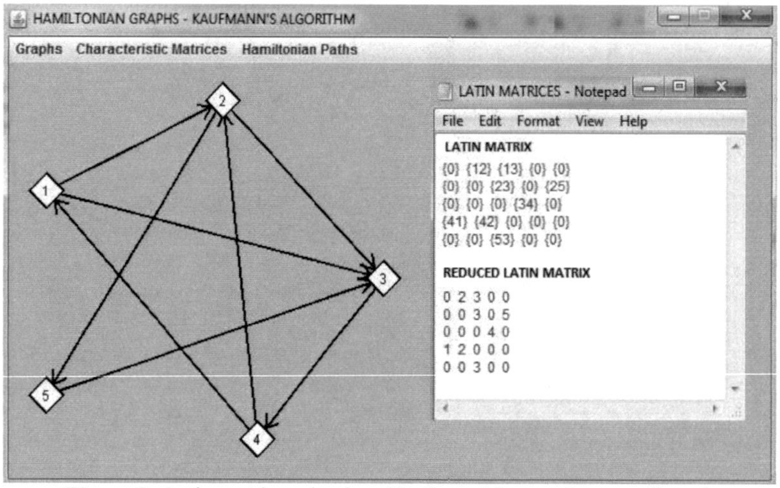


**Figure 9.** Latin matrix and reduced Latin matrix associate to a digraph

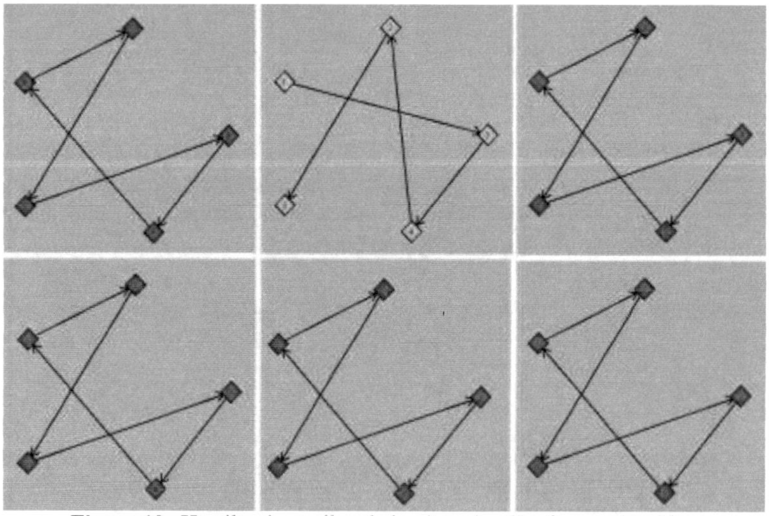

**Figure 10.** Hamiltonian trail and circuits using Kaufmann algorithm

IOP Conf. Series: Materials Science and Engineering **57** (2014) 012012 doi:10.1088/1757-899X/57/1/012012

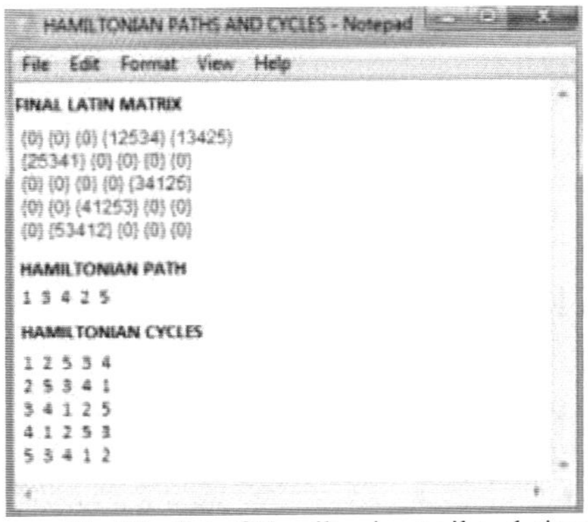


**Figure 11.** The list of Hamiltonian trail and circuits

## 4. Conclusions

The main goal in realizing of this article is designing and implementation of an interactive environment that has to lead the user to accomplish experience in comprehending and controlling acquaintances in Hamiltonian digraphs theory domain and to offer efficient and convenient access to the latest information.

The theme that is treated in this article is actually; graphs theory is an important component in developing the experience of the young mathematicians and engineers. The informatics application follows these actual problems and boards it in a new manner using modern educational technologies.

## References

[1] Chartrand G, Lesniak L and Zhang P 2010 *Graphs & Digraphs* (London: Chapman and Hall/CRC)

[2] Gould R 2003 Advances on the Hamiltonian Problem – A Survey *Graphs and Combinatorics* **19** 7

[3] Chen C and Quimpo N 1979 On some classes of Hamiltonian graphs *Southeast Asian Bulletin of Mathematics* **2** 252

[4] Kaufman A and Malgrange Y 1963 Recherche des chemins et des circuits Hamiltoniens d'un graphe *Revue Française de Recherche Opérationnelle* **26** 61

[5] Foulkes J 1960 Directed Graphs and Assembly Schedules *Proceedings of Symposia in Applied Mathematics* **10** 218

[6] Albertson M 1987 Finding Hamiltonian Cycles in Ore Graphs *Congr. Numer.* **58** 25

[7] Fowler M and Scott K 2000 *UML Distilled: A Brief Guide to the Standard Object Modeling Language* (Reading: Addison-Wesley)

[8] Dennis A, Wixom B and Tegarden D 2012 *System Analysis and Design with UML* (San Francisco: Wiley)

[9] Gomaa H 2011 *Software Modelling and Design: UML, Use Cases, Patterns and Software Architectures* (New York: Cambridge University Press)

[10] Bloch J 2008 *Effective Java* (Boston: Addison-Wesley)

# Cervical and facial infections – a real life threat

**S Rosu**[1], **M Fratila**[1]

[1] Maxillo-facial surgeon University of Medicine and Pharmacy "Victor Babeş" Timişoara, Piaţa Eftimie Murgu No. 2, 300041, Romania

E-mail: serbanrosu@yahoo.com

**Abstract.** Cervicofacial infections of dental origin are a difficult and complex issue of oral and maxillofacial surgery. Recognizing in due time the situations which are likely to develop a life-threatening condition and medical surgical prompt interventions significantly reduce the rate of the complications. Between January 2009 and March 2013, at the Clinic of Oral and Maxillofacial Surgery of the University of Medicine and Pharmacy "Victor Babes" Timisoara, 17 patients with severe cervicofacial infections were admitted in the emergency department as they needed a complex medical surgical treatment in accordance with protocol established together with the intensive-care department. Assessing the situations, we noticed a difficult, prolonged time of the recovery process which needed a hospitalization period of around 20 days. It recorded two deceased because of cervical necrotizing fasciitis and oral floor phlegmon, the most severe forms of the cervicofacial infections. The severity of the condition of the patients with cervicofacial infections must be figured and as quickly as possible an energetic therapeutic attitude must be adopted. The experience shows a frequent resistance to antibiotics like ampicillin, penicillin and oxacillin. The patients must be guided in due time to a clinic which has an intensive care department, where the surgical treatment must be administrated together with an intensive treatment for supporting the general condition. The reduction of the vital risk of the cervicofacial infections of dental origin will be done through an attentive assessment of the general and local condition (status) of the outpatients, before the dental extraction. The absence of a treatment adapted to the situation and to the clinic development, meaningfully increases the rate of the complications and the length of the hospitalization, the lethal evolution being not excluded.

## 1. Introduction

The diagnosis and the treatment of the severe cervicofacial infections present a challenging problem for the oral and maxillofacial surgeon. These infections remain an important health problem, with significant risks of morbidity and mortality, if the situations which are likely to develop a life threatening condition are not recognized in due time [9]. Because of the variety of the antibiotics, the development of new therapeutic schemes and of the safety of the surgical techniques as well as the rate of the complications of the cervicofacial infections is decreasing, especially when the medical-surgical intervention is performed in due time. The cervical necrotizing fasciitis and the phlegmon, the most severe forms, are the result of acute, diffuse infections, favored by a deficient immunologic background, having as the most frequent etiologic factor a mandibular molar tooth. The infectious process has a local expansive tendency, through the infiltration and the destruction of the cervical tissues, following the anatomic cleavage plans and an aggressive evolution with a rapid deterioration of the general condition and jeopardizing the life of the patient [7].

IOP Conf. Series: Materials Science and Engineering **57** (2014) 012013 doi:10.1088/1757-899X/57/1/012013

## 2. Objectives

The cervicofacial region presents particularities which can complicate the evolution of an infection with such a location. The anatomic complexity of the region and the deep location of the infection, render the diagnosis more difficult. The spreading of the infection is favored by the natural connection between deep neck spaces, the infection being able to exceed the limits of the region and to invade the adjacent spaces, the mediastinum and endocranial cavity. The layer of covering unaffected soft tissues can be substantial, increasing the risk of neurovascular injury, during the surgical approach.

In spite of the progress achieved by the modern antibiotics and the sophisticated diagnosis methods (MRI, CT), it is ascertained an increase of the rate of the severe cervicofacial infections, with complicated dental lesions antecedently treated through a dental extraction without a careful examination of the local and general condition (status) of the patient. Consequently, a complete assessment of the biological condition of the outpatient is necessary, as well as the recommendation of a prophylactic antibiotic treatment [8].

## 3. Material and method

Our study, completed between January 2009 and March 2013, comprises of 17 patients with cervicofacial infections of exclusive dental origin, who needed complex medical surgical procedures and intensive care treatment. The patients, 10 men and 7 women, between the ages of 18 and 65 years (an average of 40 years), came to the emergency room and were hospitalized at the Clinic of Oral and Maxillofacial Surgery in Timisoara with the following diagnosis:

- orbit abscess – 1 case
- masseter abscess – 3 cases
- submandibular abscess – 5 cases
- genian abscess – 3 cases
- oral floor phlegmon (Ludwig angina) – 2 cases
- necrotizing fasciitis – 2 cases
- laterocervical abscess – 1case

The odontogenic etiology of the cervicofacial infections include dentoalveolar infectious processes in evolution, as well as recent dental procedures:

- periapical chronic osteitis in 6 cases
- procedures involved in tooth extraction in 11 cases

Considering the initiator event, the infection of the deep cervicofacial space was produced following the ways:

- direct, by spreading along the anatomic cleavage plans and the connections between the deep cervical spaces;
- lymphatic, by spreading the infection from the oral cavity through the lymphatic system

The microbiological examination performed in these cases showed the presence of klebsiella, staphylococcus aureus, escherichia coli and pyocyanic bacillus and with variable tested sensibility to amoxicillin, cephalosporin of second and third generation (Medocef, Cefort, Rocephin) or carbapenem (Tienam), gentamicyn, ciprofloxacin, colistin and resistance to ampicillin, oxacillin, erythromycin and tetracycline[4].

The complex medical-surgical treatment was performed in cooperation with the intensive care [12] department and followed the protocol:

- Monitoring the hemodynamic parameters and the vital functions in the intensive care department [10][3]
- Combating the toxicoseptic shock through an energetic anti-infectious treatment
  Associated antibiotic therapy with wide spectrum, administrated parenteral, initially based on the clinical experience (cephalosporin in association with an aminoglycoside or quinolone with metronidazole) [6], promptly modified according to culture and sensitivity test results. If the patients are in shock before any surgical procedures, we must establish a

secure airway and in maximum emergency situations we perform cricothyrotomy and tracheal intubation through the oral passage or tracheotomy

- The prophylaxis of cavernous sinus thrombosis (heparin, clexane)[11][2]
- The surgical treatment usually applied simultaneously is meant to ensure the drainage of all the cervical spaces involved in the septic process, to aerate them and to use the antiseptics. Every approach used must be wide; most cervical infections need a transcervical approach, which facilitates an adequate exposure, with the protection of the neuromuscular structures [1]
- Anti-inflammatory and analgesic medication
- Stimulation of the general immunity through non specific vaccine therapy, administration of gamma globulins and vitamins
- The removal of the etiologic factors through extraction of the causal teeth after the local and general rehabilitation in order to gain healing and prevent relapse [5].

## 4. Results

Assessing the presented situations we noticed a difficult long-term healing process which needed a hospitalization of 10 - 40 days. A patient with cervical necrotizing fasciitis needed after 2 hours of hospitalization intubation through tracheotomy because of severe respiratory distress appeared together with toxicoseptic shock. After 8 days of continuous medical and surgical treatment, after a short improvement of the general situation, the patient died because of cardio respirator stoppage. In the other cases no relapse was noticed after the remission of the infection, the removal of etiologic factors and after discharging the patient from hospital.

## 5. Discutions

The patients with cervicofacial infections who arrived late at the Clinic of Oral and Maxillofacial Surgery and whose treatments were delayed or misconducted can expect a greater number of complications and a prolonged recovery period.

Necrotizing fasciitis, a streptococcal infection, with morbidity and mortality rate of 70-80%, was the most severe form of cervicofacial infection, as one of the two cases treated in our clinic had a lethal end. The gravity of the situation of such patients must be inferred as soon as possible and strong therapeutic measures must be taken. We have to administrate antibiotics in high doses, with a good penetration in the soft tissues and bone, based on an antibiotics with wide spectrum, like cephalosporin of the second or third generation (Medocef, Cefort, Rocephin), or carbapenem (Tienam), associated with a second antibiotic of the aminoglycoside group (gentamicin, neomycin) and metronidazole to cover also the anaerobic bacteria, present in the deep, unaerated cervicofacial deep spaces and the patient must be directed into emergency care, in the intensive care unit. It is compulsory to obtain microbial cultures from the pathological secretions, as well as blood cultures, consequently adapting the antibiotherapy according with the sensitivity tests. The experience proves frequent resistance to antibiotics like penicillin, ampicillin, and oxacillin, those being used nowadays only when their bactericide action is indicated by tests. Amoxicillin is more efficient associated with clavulanic acid (amoxyclav, augmentin ). Surgical treatment is instituted as soon as the preoperational conditions are fulfilled, accomplished simultaneously with the intensive anti-infectious treatment and supporting of the general condition.

## 6. Conclusions

The abscess of orbital, laterocervical abscesses, the hemifacial and oral floor phlegmon are the cervicofacial infections which, when the diagnosis and/or institution of an adequate treatments are delayed, can threaten the life of the patient.

Necrotizing fasciitis is a dangerous disease which mobilizes all the efforts in an attempt to save the patient's life, which is really endangered.

The occurrence of severe cervicofacial infections among young people is favoured by decreased immunity caused by some exhausting factors (stress, deficient nutrition, pollution), as well as by increased resistance of the microorganisms to the usual antibiotics with the wide spectrum.

In the patients with cervicofacial infections who do not receive treatments adapted to their situation and to their clinical evolution, complications can occur, which prolong the healing time, and this kind of lethal evolution cannot be accepted.

We can prevent the life-threatening danger of such infections through a careful assessment of the general and local condition of the outpatients and by starting a prophylactic antibiotheraphy, in accordance with therapeutic plans based on the latest clinical research.

**References**

[1] Pfanner W 1918 Fur Kenntis und Behandlung des Nekrotisier - den Erysipelas *Deutsch. Ztschr. f. Chir.* **108** 144

[2] Wilson B 1952 Necrotizing fasciitis *Am. Surg.* **18(4)** 416

[3] Klabacha M E, Stankiewicz J A and Clift S E 1982 Severe soft tissue infection of the face and neck: a classification *Laryngoscope* **92(10 Pt 1)** 1135

[4] Lindner H H 1986 The anatomy of the fasciae of the face and neck with particular reference to the spread and treatment of intraoral infections (Ludwig's) that have progressed into the adjacent facial spaces *Ann. Surg.* **204** 705

[5] Giuliano A, Lewis F Jr, Hadley K and Blaisdell F W 1977 Bacteriology of necrotizing fasciitis *Am. J. Surg.* **134** 52

[6] Corsten M J, Shamji F M, Odell P F, Frederico J A, Laframboise G G, Reid K R, Vallieres E and Matzinger F 1997 Optimal treatment of the descending necrotizing mediastinitis *Thorax* **52** 702

[7] Krenk L, Nielsen H U and Christiansen M E 2007 Necrotizing fasciitis in the head and neck region: an analysis of standard treatment effectiveness *Eur. Arch. Otorhinolaryngol* **264** 917

[8] Bansal A, Miskoff J and Lis R J 2003 Otolaryngologic critical care *Critical Care Clinics* **19** 55

[9] Quinn F B Jr Ludwig angina 1999 *Archives of Otolaryngology—Head & Neck Surgery* **125** 599

[10] Spitalnic S J and Sucov 1995 A Ludwig's angina: case report and review *The Journal of Emergency Medicine* **13** 499

[11] Freund B and Timon C 1992 Ludwig's angina: a place for steroid therapy in its management? *Oral Health* **82** 23

[12] Larawin J, Naipao S P et al 2006 Head and neck space infections *Otolaryngology—Head and Neck Surgery* **135** 889

# The importance of early detection of lip cancer risk groups

**M Fratila[1] and S Rosu[1]**

[1] Maxillo-facial surgeon, University of Medicine and Pharmacy "Victor Babeş" Timişoara, Piața Eftimie Murgu No. 2, 300041, Romania

E-mail: serbanrosu@yahoo.com

**Abstract.** Oral maxillo-facial region cancer carries major importance in the tumour pathology of the organism being characterized by a high frequency as well as by the variety of the clinical anatomical and topographic forms through which it is presented. Over 60% of labial carcinoma begins as an asymptomatic ulceration, therefore patients do not pay due attention, considering it a "rebellious thrush" and they make a specialized medical appointment in an advanced stage of the tumor. In this study we pursued the frequency of the lip cancer pathology compared to the total CMF; the distribution the lip cancer by sex and age in patients who submitted to the specialized service; the originating environment of the patient with lip cancer; the anatomical location of the lip cancer; the frequency of relapses after treatment; the presence of adenopathy in the first consultation. The study was performed at the Clinic of Cranio-Maxillofacial Surgery, University of Medicine and Pharmacy "Victor Babes" Timişoara and pursued statistical aspects of the lip cancer incidence over a period of five years (2007-2012). Pre- and postoperative patients were monitored constantly, registering in individual sheets the evolution of the disease, monitoring the relapses after treatment and the presence of adenopathy in the first consultation. As shown in the statistics made in the last five years (2007-2012), from a total of 8135 cases with CMF pathology hospitalized in the Timisoara surgery clinic, 163 cases, or 2%, were cancer of the lip. Analyzing the gender distribution shows that males represent 81% of cases while the remaining 19% were found in women. From the study of age distribution, we found that the number of cases increases with age: 153 cases over 60 years old and 58 cases between 20 - 60 years. Personal statistics from the 212 cases of cancer of the lip reveal that 143 (67%) patients were from the rural areas and 69 (33%) from urban areas. Neoplastic pathology is constantly increasing both in frequency as well as in the therapeutic problems raised. In the face and oral cavity cancer catagory, lip represents 2% of all cases and 19-25% of the total facial cancer area. Lip cancer is one of the localizations that, when detected early, can benefit from an effective therapy with high chances of healing. In order to achieve a complex treatment, interdisciplinary collaboration is required, only thus being able to determine both the therapeutic methods as well as their association.

## 1. Introduction

Oral and maxillo-facial region cancer carries major importance in the tumour pathology of the organism, being characterized by a high frequency as well as by the variety of the clinical anatomical and topographic forms through which it is presented.

Of the multiple localizations of head and neck cancer, labial carcinoma represents 50%, according to Gordon Castigliano, and 3% of the total cancer localizations [1].

Given the topographic location of the lips, malignant tumors appearing here have a number of particularities [5]:

- extreme polymorphism on the histological level and especially into the clinical plan, which slows their diagnosis;
- frequent occurrence of precancerous neoplasia on the lips, which gives an opportunity to be identified and treated correctly in the early stages of development;
- frequent exposure to potentially malignant causation factors (solar radiation, pollutants, etc.);
- the possibility of therapeutic approach through a multitude of treatment methods, more numerous as in the case of visceral cancer.

Today, despite the progress in therapy, it is unanimously acknowledged that the most effective weapon against cancer is to establish a diagnosis and treatment as early as possible.

Over 60% of labial carcinoma begins as an asymptomatic ulceration, therefore patients do not pay due attention, considering it a "rebellious thrush," and consequently they schedule a medical appointment in an advanced stage of the tumor [3].

## 2. Ojectives

In this study we pursued:

- the frequency of lip cancer pathology compared to the total CMF
- the distribution lip cancer by sex and age in patients who submitted in the specialized service
- the originating environment of the patient with lip cancer
- the anatomical location of the lip cancer
- the frequency of relapses after treatment
- the presence of adenopathy in the first consultation

## 3. Material and method

The study was performed in the Clinic of Cranio-Maxillofacial Surgery, University of Medicine and Pharmacy "Victor Babes" Timisoara and pursued statistical aspects of lip cancer incidence over a period of five years (2007-2012). Preoperative clinical and laboratory investigations were performed (blood tests, pulmonary and cardiology specialty consults) in view of general anesthesia and radio-imaging investigations (CT, MRI) isolating the tumor boundaries and eventually the extension to neighboring organs. Surgical procedure consisted of three stages: tumor removal in clinical oncology safety limits, reconstructive plastic surgery of the defect, and subsequent cervical therapeutic conduct. The surgical treatment was followed, if needed, by radiation or chemotherapy treatment and immune treatment. Pre- and postoperative patients were monitored constantly, registering in individual sheets the evolution of the disease, monitoring the relapses after treatment and the presence of adenopathy in the first consultation.

## 4. Results

As shown in the statistics made in the last five years (2007-2012), from a total of 8135 cases with maxillo-facial pathology hospitalized in the Timisoara surgery clinic, 163 cases, or 2%, were cancer of the lip.

Analyzing the gender distribution shows that males represent 81% of cases, while the remaining 19% were found in women [4]. However, some English authors, referring to smoking as a major etiological factor for the lip cancer, show that in the last 50 years damage reports relative to patient gender remained relatively constant in the context in which the number of women smokers increased significantly [7].

From the study of age distribution, we found that the number of cases increases with age: 153 cases over 60 years old, 58 cases between 20 - 60 years.

These statistics correspond to international data mentioning that the incidence increases with age, so at age 75 reaches 1/1.100 compared with 1/20.000 in the general population.

The environment has some significance in cancer of the lip; the rural population is affected at a rate of 60-90% more frequently due intensive exposure to wind, solar radiation and large temperature differences.

Personal statistics from the 212 cases of cancer of the lip reveal that 143 (67%) patients from the rural areas, and 69 (33%) from urban areas.

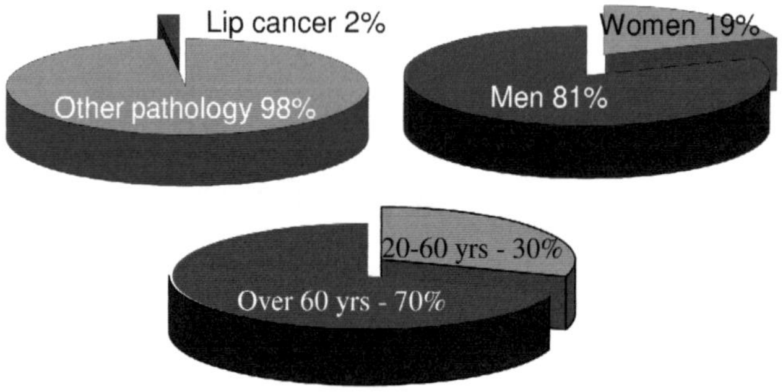


**Figure 1. a**. Prevalence; **b**. Sex distribution; **c**. Age distribution of lip cancer

## 5. Discussions

Neoplastic pathology is constantly increasing both in frequency as well as in the therapeutic problems raised. In the face and oral cavity cancer catagory, lip represents 2% of all cases and 19-25% of the total facial cancer area.

In terms of the effect on genders, males are more affected in proportion of 81% compared to 19% in women. Explanation of this difference is found in the fact that men are more frequently exposed to carcinogenic factors.

In terms of the effect on the age groups, the highest frequency is placed in the period of old age, during which the tissue changes are influenced by the senescence crisis. This does not mean that in other age groups there are no cases of lip cancer.

The highest number of lip cancer cases is found in rural areas, 67% compared to only 33% in urban areas. The explanation for this difference is the greater frequency of carcinogenic factors in the environment where the predominant agricultural workers are exposed to both sunlight as well as prolonged thermal variations and effects of the chemicals used in agriculture.

In terms of topography, 90% of cancer affects the lower lip and only 10% the upper lip, the lower lip being exposed more directly and prolonged to harmful factors.

Lip cancer receives in order of frequency: surgical care, radiant, chemotherapy and immune therapy. After treatment, the evolution of the disease can be towards post therapeutic healing (depending on the stage in which the cancer was caught) or it can be followed by treatment of local recurrence, which we found in 6% of cases.

Lip cancer treatment is not complete unless you take into account regional adenopathy [6], which was clinically assessed by us at the first consultation in 9% of cases. In the evolution of lip cancer, impairment related to spreading of the cancer to the neighboring organs, such as the mandible, may occur. These situations require a different therapeutic strategy and reduce the chance of healing.

## 6. Conclusions

Lip cancer is one of the localizations that, when detected early, can benefit from an effective therapy with high chances of healing.

In order to achieve a complex treatment, interdisciplinary collaboration is required, only thus being able to determine both the therapeutic methods as well as their association.

In early detection of the lip cancer, the dentist and the family doctor plays an important role [2]. Faced with all categories of patients, they are obliged to make a full examination of the case and to guide the patient to the cranio-maxillo-facial surgical clinic from the first signs of illness.

## References

[1] Burket L W and Castigliano S G 1965 *Oral medicine* (Philadelphia: Lippincott)
[2] Levi F, La Vecchia C, Te V C and Franceschi S. 1993 *Br. J. Cancer* **68** 1012
[3] Ifrim M and Niculescu 1998 G *Compendiu de anatomie* (Bucuresti: Editura Stiintifica si Enciclopedica)
[4] Milašin J, Pujić N, Dedović N, Nikolić Ž, Petrović V and Dimitrijević B 1994 *Journal of Oral Pathology & Medicine* **23** 298
[5] Monagham A M 1993 *British Dental Journal* **175** 419
[6] Urso C, Paglierani M and Bondi R 1993 *American Journal of Dermatopathology* **15** 435
[7] Tosios K, Koutlas I G and Papanicolaou S I 1994 *Journal of Oral and Maxilofacial Surgery* **52** 1263

# AUTHOR INDEX

**Institute of Physics**
Dirac House, Temple Back
Bristol BS1 6BE UK